The Cambridge Manuals of Science and
Literature

COAL MINING

The Alexandra Pit of the Wigan Coal and Iron Co., Ltd., Wigan

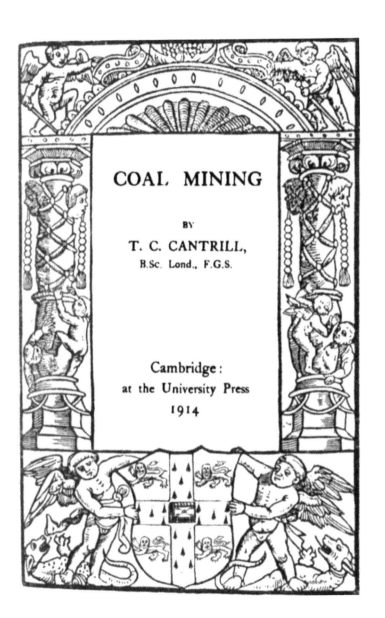

COAL MINING

BY

T. C. CANTRILL,
B.Sc. Lond., F.G.S.

Cambridge:
at the University Press
1914

CAMBRIDGE UNIVERSITY PRESS
Cambridge, New York, Melbourne, Madrid, Cape Town,
Singapore, São Paulo, Delhi, Tokyo, Mexico City

Cambridge University Press
The Edinburgh Building, Cambridge CB2 8RU, UK

Published in the United States of America by Cambridge University Press, New York

www.cambridge.org
Information on this title: www.cambridge.org/9781107605817

© Cambridge University Press 1914

First published 1914
First paperback edition 2011

A catalogue record for this publication is available from the British Library

ISBN 978-1-107-60581-7 Paperback

*With the exception of the coat of arms at
the foot, the design on the title page is a
reproduction of one used by the earliest known
Cambridge printer, John Siberch, 1521*

PREFACE

IN the following pages an attempt has been made to place before the general reader a slight sketch of the principles of Coal Mining. Not to take too narrow a view of the subject, in the earlier sections I have outlined the evolution of the industry from its primitive beginnings, and have indicated here and there some of the far-reaching effects it has had on domestic and mechanical affairs. I have also introduced such geological considerations as have a direct bearing on the main subject.

The history of Coal Mining in Britain has been written by Mr R. L. Galloway in some fascinating volumes to which I am indebted for the particulars in Chapter I. In the section dealing with leases and royalties I have had the help of Mr H. J. Randall, of Bridgend, who is conversant with the customs obtaining in South Wales and elsewhere. In the Bibliography is given a list of works laid under contribution for the present purpose, and to them the reader is referred for fuller details. The Frontispiece has been kindly supplied by the Wigan Coal and Iron Company, Ltd.

T. C. CANTRILL.

22 *December* 1913

TO

DANIEL JONES, Esq., J.P.,
OF DONINGTON, ALBRIGHTON, SALOP;
FATHER OF WYRE FOREST GEOLOGY

CONTENTS

LIST OF ILLUSTRATIONS

Note.—The Frontispiece shows, on the left, the screens with the winding-shaft (which is also the downcast air-shaft) behind them. The upcast air-shaft and fan-house are not included in the view. The head-gear on the right belongs to a disused pit to the shallower seams.

CHAPTER I

INTRODUCTION AND HISTORICAL REVIEW

Introduction.—The intimate dependence of our comfort on a supply of cheap coal was brought home very forcibly to most of us during the strike of April, 1912, when ' stove-nuts ' were quoted on the London Coal Exchange at 40 shillings the ton. Whether we use the coal itself in our sitting-rooms and kitchens, or warm ourselves, cook our food, and light our rooms with gas, we depend ultimately on the same fuel. Nor do we become independent of it by the adoption of electricity—generated as it is in most cases by steam-power raised by the combustion of coal. Our railways too, whether steam or electric, equally draw their vitality from a regular supply of the same source of energy. With our coal-supply cut off, our water-service, pumped by steam-power from the well to the reservoir, would soon fail us ; and worse things would soon befall those towns whose sewage-system depends for its proper working on the assistance of steam.

The vast amount of coal demanded by our various manufactures is not so easily appreciated ; but when we reflect that, summer and winter alike, thousands of furnaces, forges, steam-engines, gas-works and coke-ovens, brick-works and lime-kilns, are devouring the fuel without cessation, while steam-vessels are not only consuming it but are carrying it to all parts of the world for foreign consumption, we gain some notion of the extent and importance of this great British industry.

In the sequel we shall see that the requirements of the coal-trade gave origin to an important series of useful inventions. The first steam-engines were constructed for no other purpose than the pumping of water from the mines ; the locomotive was produced in order to convey the coal from the pit to the port of shipment, and with the introduction of iron rails laid the foundation of our present railway-system. In fine, the domestic, the municipal, and the commercial life of modern Britain depends for its very existence on, as it derives its vigour from, the fortunate circumstance that, many millions of years ago, some of the forests and swamps of the Carboniferous period spread across the site of the future Britain.

Historical Review.—Though the use of coal or lignite by the smiths of Liguria and Elis (Genoa and Southern Greece) is recorded by Theophrastus about 300 B.C., there is no evidence that the mineral was known in Britain before the Roman occupation. The

abundant supply of timber sufficed for all the needs
of the natives, who required no lime in the construction of their primitive dwellings, and smelted their
bronze and iron with wood or charcoal. There is
little doubt, however, that it was employed to some
extent by the Roman colonists, for smith-work and
lime-burning. They also employed it for heating-
purposes, on occasion, for coal-cinders were found in
plenty in the hypocausts at Uriconium (Wroxeter)
in Shropshire, and coal or its cinders have been discovered on the sites of many of the forts along the
Wall of Hadrian. Their use of it seems however to
have been very limited; no Roman remains have
been discovered in any of our coal-workings, and
though in the north of England they built their mili-
tary stations close to the outcrops of the coal-seams,
the Romans appear to have left the coal practically
untouched.

The Saxon and English invaders seem to have
known nothing whatever about the mineral. To
them wood was the all-sufficing fuel; what little
iron they had was smelted with charcoal, and their
buildings, with the exception of a few churches, were
constructed of timber, and needed no mortar; any
lime they used was doubtless burnt with wood. They
warmed their halls and their hovels alike with wood
and peat, even in districts that abounded with coal.
In Domesday Book no mention is made of coal,

though other minerals are alluded to. It could not
have been long, however, before the Norman builders
of castles and religious houses began to burn their
lime and forge their iron with coal, but there is great
difficulty in adducing contemporary records as evidence of this, owing to the fact that originally the
term 'coal,' or, as formerly spelt, 'cole,' like the
Greek *anthrax* and the Latin *carbo*, signified any fuel,
generally wood. Unless therefore the document ap-
pealed to contains some contextual allusion to a pit,
it is impossible to assert that the passage in ques-
tion refers to the mineral fuel. Similarly the term
'collier' meant at first 'charcoal-burner'; and the
'Wood-collier's Arms' still survives (or did in 1895)
as the name of an inn at Bewdley, affording an
instance of this usage of the word among the char-
coal-burners of the neighbouring Forest of Wyre.

There appears to be no uncertainty however
about the records of Holyrood and Newbattle Abbeys, which allude to the digging of coal on the
south shores of the Firth of Forth about the year
1200 ; and early in the reign of Henry III coal began
to be gathered along the coast of Northumberland,
where it was washed up by the surge from outcrops
on the shore, and thus acquired the distinctive name
of 'sea-coal'; and what is perhaps the first unequi-
vocal reference to the mineral in England is contained in a grant, to the monks of Newminster

Abbey, by Adam de Camhous, of land on the coast near Blyth, with a road to the shore for the conveyance of sea-weed and sea-coal (*carbo maris*). This was a few years prior to 1236. With regard to the term 'sea-coal,' it is of interest to find that by the time of Henry VIII the origin of the name had become a matter of uncertainty ; Leland regarding it as derived from the fact that the mineral was gathered on the shore, while Dr Caius attributed it to the mode in which the coal was conveyed to London.

During the reigns of Henry III and Edward I, coal-digging sprang up in most of the coalfields, but was most active in the great northern coalfield (Northumberland and Durham), owing to the facility with which the mineral could be floated down-stream to the coast at Tynemouth. It was not long before it began to be shipped thence to London, where as early as 1228 it appears to have been sold to the lime-burners of Sea-coal Lane (still in existence near Ludgate Circus) ; and as one William of Plessey had property in Sea-coal Lane in 1253, the village of Plessey (north of Newcastle-on-Tyne) was probably the source of the first coal to reach the metropolis. In 1257–9 ship-loads of sea-coal arrived in London for the smiths—and lime-burners, probably—at work on Westminster Palace. In London the brewers and dyers were using it in 1306,

though it aroused the opposition of the citizens on
account of its noisome smoke.   Coal was employed
by the smiths and lime-burners engaged on the
Edwardian castles about 1300, *e.g.* Carnarvon,
Beaumaris and Dunstanborough, as can be gathered
from contemporary works-accounts ; and in 1366–7
some 576 tons of it were brought from Winlaton in
Durham county for works at Windsor Castle.

About 1300–25 coal began to be tried in a very
shy fashion in the castles, abbeys and better sort of
houses ; for improvements in architecture carried
with them improved chimneys and fireplaces, without which the new fuel, with its rank smoke, could
hardly have displaced the less sooty and pungent
wood-fire from the central hearth.   By the middle
of the 14th century the general demand for coal had
increased considerably, and as early as 1325 a boat-
load of the mineral left Newcastle for Pontoise in
France; but this foreign exportation was prohibited
in 1362 and 1367, except to Calais.

Up to this time the getting of the coal was not
a very arduous business.   The mineral no doubt
was obtained at first from the actual outcrop, *i.e.*
from the tract along which the coal-seam lay immediately below the soil, and could be got by simple
quarrying.   This method of ' open-work ' or ' open-
cast ' would be specially applicable in those districts
where the coals crop out along the steep sides of

hills and valleys (Fig. 1, p. 9). In such situations, moreover, the coal could readily be followed underground from its outcrop, and worked by horizontal tunnels known as ' day-holes ' or ' day-levels,' which served the double purpose of affording an exit for the coal and allowing the works to drain themselves. But these modes were less suitable in flatter districts, such as parts of South Staffordshire ; and there resort was had to the sinking of ' bell-pits ' or ' beehive pits.' These were shallow pits sunk through the surface-beds to the desired seam of coal (or of ironstone), upon reaching which the pit was belled-out, and as much of the mineral removed as could be done with safety. The pit was then abandoned, and filled up with refuse from a new pit sunk hard by.

But by the middle of the 14th century opportunities for the application of these simple methods were becoming fewer in the north of England, and we begin to read of pits and water-adits, ropes and windlasses ; in fact, coal-mining had entered on the second stage of its evolution, the ' pit-and-adit ' stage (Fig. 1, p. 9). The earliest mention of coal-mining implements occurs in an inventory dated 1354 of property belonging to the monks of Finchale (on the Wear), in which are included *ij colpikkes*, *ij yeges ferrei*, *i.e.* two coal-picks and two iron wedges.

During the latter half of the 14th century the use of coal extended rapidly for all manner of

purposes where wood was employed before. The
monks of Holy Island were using it in 1344-7 for
warming their hall, their prior's chamber, and their
infirmary, as well as in their brew-house and their lime-
kiln. It was necessary now to win the coal over areas
farther removed from the outcrop, and to follow it
down in the direction of the dip. Pits were therefore required for raising the coal; and to allow the
workings the benefit of natural or 'free' drainage,
long narrow tunnels (adits, soughs, water-gates) were
driven up to the workings from the lowest valley-
bottom available, an arrangement that also provided
the workings with a natural ventilation. The coal
was carried to the bottom of the pit, or out of the
level, on the backs of boys, girls and women, known
as 'bearers,' and was raised to the surface in baskets
with hempen ropes and windlasses.

During the 15th century the use of coal was
steadily spreading. In London it was taking the
place of wood on the hearths of the citizens, and in
the maritime regions it was coming into use in the
evaporation of sea-water for the manufacture of
salt. Mention of water-gates or adits becomes more
frequent, indicating that in many districts the pits
were being deepened; and towards the end of the
century (1486-7) the monks of Finchale had been
obliged to set up a pump at their pits, which had
apparently passed below the level of natural drainage,

and had entered on the third stage of their evolution, viz. the ' pit ' stage (Fig. 1, below), when it became necessary to raise both coal and water by artificial means. In many of the coalfields, however, the ' pit ' stage was not reached till the close of the 16th century. The tools used at this time were few and simple : picks and wooden shovels, 'scopes' (probably buckets) and ropes were all that were needed. At

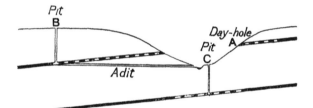

Fig. 1.—Section showing the three stages of coal-mining. *A*, the day-hole or day-level ; *B*, the pit-and-adit ; *C*, the pit. In the first, no machinery is needed for haulage or drainage ; in the second, the coal is raised up the pit by machinery, and the water drains away by the adit ; in the third, both coal and water are raised by machinery. (After R. L. Galloway.)

the close of the century the pitmen were still more or less serfs, and in some districts continued so till the reign of Elizabeth, who freed some of her serfs in 1574.

In the 16th century the growing scarcity of wood, which was steadily disappearing into the furnaces of the iron-smelters and salt-makers, gave an impetus

to the use of coal for domestic purposes. This was
facilitated by improvements in the construction
of fireplaces and chimneys that came in about the
middle of the century. Various Acts of Parliament
for preserving the woodlands and restraining the
activities of the iron-makers were passed in the
reigns of Henry VIII and Elizabeth, but with little
effect. In the Newcastle district the general domestic employment of coal appears to have begun
about 1570, previous to which its use seems not to
have extended beyond the bloomary, the smithy
and the lime-kiln ; and by the middle of the century
a considerable foreign export had grown up ; but in
view of the feared scarcity of fuel this trade was not
encouraged. The corf or circular hazel-rod basket,
in which the coal was drawn up the pits, is first
mentioned in 1539 ; it was provided with a wooden
bow for attachment to a hook at the end of the rope.
This primitive vessel continued in use in some districts for special reasons even as late as 1871 !

About the middle of the 16th century coal was
being used largely for salt-making by the monasteries
along the coasts of Northumberland and Durham ;
and in 1555 we first meet with a reference (in a book
by Dr John Caius, one of the founders of Gonville
and Caius College, Cambridge) to the noxious vapours
given off during the working of the coal ; and the
first recorded underground fire burned for some years

at Coleorton in Leicestershire in the reign of Henry
VIII. There was still a strong objection on the part
of the fine ladies of the metropolis to the domestic
use of coal, on account of its sulphurous smoke and
smell. It may have been this that prompted the
first attempts to make coke ; certain it is that John
Thornborough, Dean of York, was granted a patent
for that purpose in 1590. In this century too the
idea of smelting metals with coal instead of with
wood and charcoal began to exercise men's minds
and several patents were granted, but the schemes
all came to nought. Coal was slowly driving wood
from the kitchen, the hall and the salt-pan, but not
till nearly two centuries later did it force its way
into the smelting-house.

With the opening of the 17th century, we find
that James I, however much he may have objected
to tobacco-smoke, had no prejudices against that of
coal, a fuel he used in his own chamber. Coal now
came rapidly into general domestic use, except in
districts remote from the coalfields, and even there
the free use of wood was looked upon as an extrava-
gance. Still, the total vend of the northern ports
in 1609 was only 251,764 tons, of which a little over
a tenth was sent abroad. In that district fears were
already arising that the exhaustion of the mines was
not far off, as the water could be kept down only by
a continuous and desperate struggle. In 1658 is

first recorded the breaking-in of water from old workings, with fatal results, at Newcastle ; and the first
noticed death by an explosion of firedamp took place
at Gateshead in 1621.

From the beginning of this century various
schemes were patented for substituting coal for wood
and charcoal in some of the manufactures. One of
the first of these had to do with glass-making, which
previously had been seated amid the woodlands
of Sussex. After repeated failures, Robert Mansell,
Vice-Admiral of England, succeeded in establishing
the new process at Newcastle-on-Tyne in 1619, and by
1624 the works employed 4000 hands. The impetus
thus given to the manufacture of glass soon made
itself felt in the increased number of windows introduced into domestic buildings.

It is in the beginning of this century that the
practice of boring for coal is first heard of. Boring-
rods appear to have been made known in the north
of England by one Master Beaumont about 1610,
and were used in many cases where it was deemed
advisable to prove the ground before sinking a pit.
Another notable innovation about the middle of the
century was the construction of railways, which had
been in use for many years in the mines of Germany.
The wooden wheels of the wagons were flanged and
ran on wooden rails.

Draining the mines now became the most pressing

care of the proprietors. In many districts the workings were by this time below the level of free drainage,
and the mines were dependent on mechanical means
for raising the water. Chain-pumps, introduced apparently from Germany, were used about 1670 in
the north, and were actuated by horses or water-
wheels; they consisted either of an endless chain of
buckets similar to the present-day river-dredger, or
of a chain of discs passing up a tube, as in the pumps
frequently to be seen in our farmyards. By having
several special pits of graduated depths, water could
by these means be raised from depths of 240 feet.
Where the workings had not yet got below the level
of free drainage, long adits were cut with the pick
at great cost from the lowest valley-bottom available; some of them being several miles long and
only 18 inches wide.

At this time the coal was raised by horizontal
horse-gins of the 'cog-and-rung' pattern. This
machine consisted of a drum or barrel (on which
the rope was wound), placed in a horizontal position
over the pit-mouth. One end of the barrel was of
smaller diameter and built of bars or rungs, with
which the upright cogs of a horizontal wheel on a
vertical axle were made to engage. This cogged
wheel was driven round by a long arm to which the
horses were harnessed. The workings were ventilated by the air-currents naturally set up by the

provision of two or more shafts, or a shaft and water-
adit.

In 1708 was published the first treatise on coal-
mining as practised in the Sunderland and Newcastle
districts.   In *The Compleat Collier*, the anonymous
author, ' F. C.,' has preserved for us a clear picture
of the methods in vogue at that time in the north
of England.   From it we learn that the upper parts
of the pits were usually square, 6 feet in the side,
but sometimes only 3 or 4 feet, and lined with timber.
In wet strata a water-tight lining was employed,
resembling the staves of a cask and called tubbing.
Pits of 300 or 400 feet were regarded as exceptionally
deep ; the usual depth was 120 to 180 feet, the cost
of such a pit being £55.   At the working-places the
coal was filled into corves, several of which were
then placed on a wooden sledge and dragged by the
' putters ' to the bottom of the shaft, then hooked
to the rope, and drawn to the surface by the horse-
gin.   Each corf held about 4½ cwt., and the hempen
rope was about an inch in diameter.   The gin now
employed for raising coal and water was an improvement on the earlier ' cog-and-rung ' gin, and was of
the vertical whim-gin pattern (Fig. 2, p. 15).   The
rope-roll was a large wooden drum, with its axle
vertical, and was driven by horses harnessed to long
levers.   From the drum the rope passed over a
pulley in a frame built over the pit-mouth.   This

arrangement allowed the use of more horses and a larger drum than in the earlier form of gin, and did not require the mechanism to be erected so close to the pit-mouth. Ventilation was attained by natural currents from one shaft to another, the air being directed along the working face only—a method known as face-airing. The coal was got by the ' bord-and-pillar ' system (*i.e.* the coal was in part excavated, and in part left as pillars), and in a pit only 360 feet deep more than half the coal was left behind to support the roof. The practice of removing any part of the pillars was evidently unknown in the north in 1708. Thus far ' F. C.'

Underground fires were very prevalent at

Fig. 2.—A South Staffordshire horse-gin. (From a photograph by John Rhodes, junr., in the Geological Survey Collection of Photographs.)

this period, as at Pensnett Chase and near Wednesbury, both in South Staffordshire. Dr Plot in 1686 makes one of the earliest references to the use of the fire-lamp as practised at Cheadle in the same county for ventilating a mine ; it consisted of a large iron brazier of burning coal, suspended in one of the shafts (the 'upcast' shaft), which caused an upward current of air, the corresponding downward current descending the 'downcast' shaft. In the adjacent county of Worcester, Dud Dudley was devoting heroic efforts to unite the coal and the iron industries of his native district of Dudley; but although he succeeded in 1620 in producing a certain quantity of iron by his new method of smelting the ore with coal in lieu of charcoal, his repeated attempts to introduce the process on a commercial scale were defeated by the vested interests of the charcoal-iron makers.

In 1675 we read of colliers being shot out of a pit's mouth by an explosion at Mostyn; and the earliest account of the practice of deliberately setting fire to the firedamp in order to get rid of it is contained in a paper relating to the same place and communicated to the Royal Society in 1677.

The implements hitherto employed by the miner in the work of excavation were the pick, the wedge, and the hammer; but in the rare cases where these did not suffice, the ancient process known as 'fire-setting' was adopted. The rock to be removed was heated with fire and then suddenly cooled with water, with the result that the rock was cracked and fissured, and could then be cleared away with the pick and shovel. But about 1719 gunpowder, which for nearly a century had been employed in the German mines, was adopted in sinking pits in Somerset, though not for blasting the coal itself till 1813, when it was resorted to in the north of England.

At the beginning of the 18th century the greatest depth to which the pits had attained was about 400 feet. They now began to enter the dry belt of strata that underlies the watery zone, and it was not long before serious explosions took place in the fiery northern coalfield. The first happened at Gateshead in October, 1705, and involved the deaths of over 30 people, one unfortunate youth, Robert Broune, being blown up a 67-fathom shaft and flung out of its mouth!

But the event that above all others marks out this period as one of transcendent import to the mining industry—and, as it subsequently proved, to mankind in general—was the invention of the atmospheric engine. Many of the collieries had been drowned out and abandoned; the appliances in use for draining them had reached the limit of their powers; and unless some new device could be hit upon, it seemed likely that many coal-districts would be compelled to close down. At this juncture Thomas Newcomen, ironmonger, and John Cawley, plumber, of Dartmouth, anabaptists, succeeded in 1710 in giving practical shape to the ideas of Papin and other physicists, who had demonstrated that the pressure of the atmosphere will depress a piston in a cylinder when a vacuum has been produced beneath the piston. The result was the atmospheric or fire-engine, to which at first was assigned no other part than the pumping of mines. Newcomen carried out the idea by filling a vertical cylinder (open at the upper end) with low-pressure steam led from a boiler, then condensing the steam in the cylinder by an injection of cold water, and allowing the resultant atmospheric pressure to push down the piston. The piston was attached by a chain to one end of a horizontal beam (a giant pump-handle, in fact), pivoted at the middle, to the other end of which was attached the pump-rod, which was weighted sufficiently to raise the piston after its downward stroke. But as Thomas Savery had already secured a patent (from 1699 to 1734) for raising water from mines by the agency of fire, Newcomen's engine had to be produced under Savery's patent, though the two machines had nothing in common. By Savery's machine—a sort of 'pulsometer pump'—water could be raised some

26 feet by suction (produced by the condensation of
steam), and then forced up a further 64 feet by the
action of high-pressure steam applied directly to the
water. So far however as draining mines was concerned, the affair proved worthless. Several were
erected that served to supply gentlemen's mansions
with water, but the only one that essayed to drain
a flooded pit (near Wednesbury in South Staffordshire) tore itself to pieces !

The first of Newcomen's engines was erected in
1712 near Walsall in South Staffordshire ; and another, built about 1713 at Griff, near Nuneaton in
Warwickshire, at once saved £650 a year in the cost
of horses. Thenceforward the employment of Newcomen's engines for pumping mines rapidly extended
into other coalfields—particularly where water was
not available for driving water-wheels and chain-
pumps—and rendered it possible to work coal-seams
previously 'drowned.'

In the middle of the 18th century iron was still
scarce, and entered to quite a small extent into the
construction of the fire-engines, railways and winding-gear of the collieries ; the engines were built of
copper, brass and lead ; wooden rails (as the name
implies) were used on the railways, and the winding-
ropes were of hemp. But at last the grand alliance
between the coal and the iron industries was brought
about at the furnaces of Coalbrookdale Foundry in

Shropshire somewhere about 1730, and the dreams of Dud Dudley were at length realized : iron was smelted on a commercial scale with coal. Yet so quietly was this revolution brought about that the actual date seems to have escaped all record ; but it is probable that Abraham Darby began to employ coke at Coalbrookdale between 1730 and 1735, and at first may have used it mixed with charcoal, a fuel that had not wholly been abandoned there in 1803. Under the influence of the new process a rapid revival of the iron industry set in ; the cheapened metal soon made its way to the collieries, and was adopted for their equipment and machinery ; and a large number of fire-engines with iron cylinders cast at Coalbrookdale were built between 1750 and 1775 in all parts of the country.

About the middle of the century the deepening of the collieries, rendered possible by the improved methods of drainage by the new atmospheric engines, gave rise to several fresh difficulties. The raising of the coal by the old horse-gins was becoming a tedious business ; underground haulage was growing more and more costly, now that the workings were carried to greater distances from the shafts ; and the standard of ventilation that had obtained hitherto was proving inadequate for the more extensive, drier, and fiery workings now being dealt with. In spite of improvements the horse-gins were still inefficient;

and Michael Meinzies about 1750 introduced the balance-tub, an arrangement whereby the descent of a bucket of water was made to raise the coal up the shaft. But this plan could usefully be adopted only where the water thus carried into the workings could drain away by an adit. The same principle underlay his self-acting inclined planes, which enabled the full tubs to pull the empties up the underground inclines. Carriages drawn along wooden railways by horses now began to be substituted in the workings for the sledges dragged by boys. Instead of the whole of the air-current being carried past the face where the coal was being worked ('face-airing'), it was now directed, by an arrangement of trap-doors and stoppings, through every part of the excavations—a system known as 'coursing the air.' In order to avoid explosions, the working-places in the fiery mines of the north now began to be illuminated with the steel-mill, invented by Carlisle Spedding of Whitehaven about 1750. The machine consisted of a small disc of steel which by suitable gearing could be rotated rapidly by hand against a piece of flint, and so caused to emit a stream of luminous sparks. Unfortunately it was not till after a number of explosions had occurred that the steel-mill was found to be not much safer than the tallow candle. In 1753 cast-iron flanged wagon-wheels began to be substituted for the wooden ones, and in 1767 plates

of the same material were used in the Coalbrookdale
district to arm the wooden rails.

A great step forward in the evolution of the
atmospheric engine was taken in 1769 when James
Watt introduced the separate condenser and other
improvements, and applied them to the Newcomen
engines, in which the cylinder, as we have seen (p. 18),
was originally used as cylinder and as condenser in
turn, an arrangement involving great loss of steam
and waste of fuel. The simpler engines on Newcomen's original model continued in use, however, for
draining coal-mines, where suitable fuel was cheap.
Watt's engine was moreover still a single-acting low-
pressure atmospheric engine. Attempts to adapt it
to the raising of the coal met with little success ;
and this work continued to be performed in most
districts by horse-gins of horizontal or of vertical
patterns up to 1777. But about this time water-
wheels began to be employed for raising the coal,
especially where natural streams were available. One
pattern had a double set of buckets opening in opposite directions around the periphery, so that the
motion could be reversed when the full corves reached
the top of the shaft. Such a wheel was erected by
Smeaton at Griff near Nuneaton about 1774 for drawing coal and water. Where a natural stream was
not available, an atmospheric engine was in some
cases actually employed to pump water up to a

cistern for the special purpose of supplying the water-wheel ! But soon afterward (in 1779) Matthew Was-brough of Bristol patented an atmospheric engine designed to produce rotary motion from a reciprocating one by means of ratchet-wheels with the addition of a fly-wheel ; but the mechanism frequently getting out of order, Watt introduced the simple device of the crank. It is said that this idea was pirated by one of Wasbrough's party ; certain it is that much to Watt's annoyance it was patented by one James Pickard of Birmingham in 1780, with the result that to obtain his rotary motion Watt fell back on the sun-and-planet mechanism. The engine was still a single-acting low-pressure engine ; but in 1782 Watt once more came forward with one of his masterly inventions by arranging to produce a vacuum alternately above and below the piston, and so made it double-acting and capable of working equally well in both directions ; at last the piston could push as well as pull ; and later, by excluding the atmospheric pressure entirely, Watt converted the atmospheric engine into the low-pressure condensing steam-engine. At once it entered upon a wider field of application, though curiously enough it still retained, even when applied to widely different ends, the badge of its original servitude, viz. the pump-handle or beam.

The production of iron with coal-fuel at Coalbrookdale had led to the erection of numerous blast-furnaces in other coalfields, so that the metal soon became plentiful; and by 1788, out of the year's total make of 68,300 tons of iron, the coke-made metal accounted for 53,800, Shropshire leading with 23,100 tons from 21 furnaces. It was natural that the first iron railways should have been made about this time in Shropshire. Cast-iron tubbing for lining the shafts was now introduced, and a number of other improvements in mining practice mark the last decade of the 18th century. Atmospheric engines were used to raise the coal ; a general substitution of iron for wood on the surface-railways took place ; and self-acting inclined planes were installed both above and below ground. Underground railways of wood became general, on which a wheeled rolley or carriage capable of holding several corves was drawn by a horse. The practice of leaving larger pillars with a view to a second working began to be adopted, as up to this time, in the deep pits on the Tyne, $45\frac{1}{2}$ per cent. was the greatest quantity of coal thought to be obtainable. A most important invention was made at this time by John Curr of Sheffield, who in 1788 introduced ' guides,' *i.e.* wooden grooves, carried up each side of the shaft, for the reception of the ends of the cross-bar to which the loaded carriages were suspended. In this way he secured smoothness and freedom from collision during the winding. He next introduced, both above and below ground, light cast-iron railways of the plate-rail type, *i.e.* having the flange on the rail and not on the wheel (Fig. 3, below). He invented the flat rope also, which could be rolled upon itself like a Catherine-wheel, and so made to equalize the work of the winding-engine.

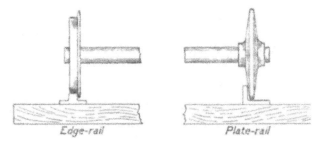

Fig. 3.—Edge-rail and plate-rail. In the first, the wheel is flanged and runs on the edge of the rail; in the second, the wheel is not flanged, but runs on a flanged plate. Our modern railways and tramways are of the edge-rail type.

In the South Staffordshire Thick or Ten-yard Seam, which on account of its liability to spontaneous combustion was worked with a minimum of ventilation, the inflammable gas was even at this time (1798) got rid of by the process of 'firing' or deliberately exploding it. The operation, a somewhat hazardous and exciting one, was carried out by the 'firemen,' who obtained their title from being told off to do this special work. At Lord Dudley's

pits at Netherton it was the practice to fire the gas
three times a day. The *modus operandi* was as
follows. The place in which the gas had accumulated having been ascertained beforehand, the
firemen proceeded from some distant side-chamber
(usually the underground stables) toward the gas-
eous part of the mine, paying out a thin copper wire
as they went. Approaching as near to the danger-
place as was prudent, they passed the wire over a
pulley at the end of a long pole, which they then
raised aloft into the gas and fixed securely in the
required position. To the free end of the wire rest-
ing on the ground they then fastened a lighted
candle, so weighted as to keep itself upright and
steady. Retiring now to their place of retreat (the
stables), they barricaded themselves in, and began
hauling-in the wire. This raised the candle at the
other end into the gas, which exploded with great
violence. The excitement arose when from some
unforeseen cause the explosion failed to come off,
so that the firemen, like boys with a toy cannon,
were faced with the alternative of venturing out to
investigate, and possibly to be caught by the explosion, or of remaining prisoners in their own castle.

At Whitehaven, Carlisle Spedding tapped the gas
issuing from fissures in the coal and conducted it in
pipes to the surface; he even proposed to light the
town with it, but doubtless the worthy citizens

were shy of having any dealings with such a dangerous illuminant.

In 1794 the ' longwall ' way of working the coal,
by which all of it was removed at one operation and
no pillars left, was being applied to some of the thin
seams in the Cumberland coalfield, though it had
long been used in Shropshire, to which county it
appears to be indigenous.

The dawn of the 19th century, owing to the
lapsing of Watt's patent rights in 1800, saw a rapid
extension of the use of steam-power. This immediately created an increase in the demand for coal
as fuel, and the extension of the coal-iron manufacture had the same effect. In the early years of
the century gas-lighting was introduced by William
Murdoch at Boulton and Watt's works at Soho,
Birmingham, in 1802, and by F. A. Winsor in London
shortly afterward, thus making a new call on the
coalfields. In 1800 the estimated coal-output of
the United Kingdom was 10 million tons.

The most notable improvements, connected with
the collieries, that mark the early part of the century
are concerned with traction. Cast-iron railways both
above and below ground came in rapidly in the
colliery-districts, though at first horses still supplied
the motive-power; wrought-iron rails did not become
usual till 1820 or thereabout. Steam-engines for
pumping and winding were becoming general, and

stationary engines were being set up for drawing
wagons up inclined railways by means of ropes. But
the improvement that ranks highest at this time was
made by Richard Trevithick, who simplified Watt's
engine by getting rid of its condenser and supplying
it with high-pressure steam, and thus converted it
into the high-pressure non-condensing engine, or
' puffer.' Already Murdoch in 1784 had constructed
a model locomotive engine, but carried the idea no
farther. In 1802 Trevithick, in association with
Andrew Vivian, patented certain improvements in
the steam-engine and their application to the pro-
pulsion of carriages, and in 1803–4 built at Pen-y-
daren near Merthyr Tydfil a locomotive that successfully drew 5 wagons carrying 10 tons of iron and
70 men a distance of 9 miles, though the strength
of the permanent way proved insufficient to bear
the load. But Trevithick somehow failed to popu-
larize his engines, either in South Wales or in the
north of England ; and although in 1812 John
Blenkinsop, by discarding horse-traction in favour
of engines of his own design, had been able, at Leeds,
to reduce the cost of haulage to one-sixth of its
previous figure, it was not till 1829 that Robert
Stephenson, by a happy combination of improvements introduced by others, built the Rocket and
placed the locomotive-engine on a sound commercial
footing.

At this time the ventilation of the mines continued to be produced by fire-lamps, or by furnaces placed at the bottom of the upcast shaft; and in fiery places in the workings, light was still obtained from the steel-mill. But a long list of explosions in the north marks the beginning of the 19th century, and shows that the ventilation was inefficient, and the steel-mill—and even the ventilating-furnace itself—a source of danger. The great problem of the time was : how to work the mines without risk of explosion. In some cases the gas was carried in pipes to the surface and burnt there; and in 1805 James Ryan proposed by taking advantage of the low specific gravity of the gas to drain it off by special passages; and in 1808 he successfully applied his system to the Netherton Colliery near Dudley.

Another difficulty that had to be contended with, especially in the deep pits of the north, was ' creep.' This phenomenon consists of a bulging-up of the floor of the excavated passages, and its ultimate coalescence with the roof. It arises where the pillars of coal left to support the roof are of insufficient size in proportion to the passages, so that the whole weight of the overlying strata, being thrown on to supports too small to carry it, forces the pillars downward into the floor-strata—usually soft and yielding—with the result that these buckle upward wherever free to do so. At the same time the pillars

themselves are frequently crushed, the coal therein
spoilt, and the workings thrown into a state of
dangerous insecurity.   And as creep set up at one
spot necessarily throws more weight on adjacent
pillars, the disease is apt to spread rapidly throughout the whole colliery.   To guard against this disaster, John Buddle, junr. in 1810 introduced ' panel-
work,' *i.e.* the laying-out of the workings in districts
or panels (? like the panels on a door), of 30 acres or
more, separated by wide barriers of solid coal 40 to
60 yards wide.   Thus if crush and creep appeared
in one panel, they might be prevented from spreading
into those adjacent.

While developing the system of panel-working
at Wallsend, Buddle introduced his compound or
' split-air ' method of ventilation.   Instead of the
whole current from the downcast shaft being carried
along every passage in the mine (a distance in some
cases of 30 miles), he divided it at the bottom of
the shaft into two or more currents, each of which
traversed only one panel.   But though Buddle's
method, which still remains the most efficient system
of air-distribution, did much to render the atmosphere of the mine safer and more wholesome, it was
powerless to prevent risk of explosion from sudden
discharges (' blowers ') of gas, either from old workings or from fissures in the coal itself, so long as
naked lights were employed.

At this juncture Dr Clanny of Sunderland devised a lamp that could be used in an explosive
atmosphere without firing it, but the instrument
was not convenient for practical purposes. About
the same time an explosion near Jarrow in 1812
determined the incumbent of the parish, the Rev.
John Hodgson, to employ his pen in bringing the
facts before the general public. This led a London
barrister, J. J. Wilkinson, to form a society for the
prevention of accidents in coal-mines, and in 1815
this body obtained the help of Sir Humphry Davy
in the matter. That philosopher discovered that
a lamp furnished with sufficiently small air-holes
would not communicate flame to an explosive
mixture outside ; and finally he produced the wire-
gauze lamp, the " metallic tissue permeable to light
and air and impermeable to flame." Davy nobly re-
fused to patent his invention, preferring not to en-
hance the cost of an instrument designed to preserve
the life of man. His safety-lamp not only did this,
but also enabled immense quantities of coal to be
got that otherwise would have been, and actually
had been, abandoned ; and by the removal of the
pillars, rendered possible by the Davy-lamp, 80 per
cent. of the coal could now be got out.

In the first quarter of the 19th century the shafts
in the north varied from 6 to 15 feet in diameter ;
and at Monkwearmouth had attained a depth of

1590 feet, the cost of a single shaft in some cases reaching as much as £40,000. In Wales and the Forest of Dene, on account of the depths of the valleys, free natural drainage was still available even in 1835. By 1827 Shropshire had been outstripped by both Staffordshire and South Wales in the make of pig-iron, the latter leading with 272,000 tons, while Staffordshire produced 216,000 and Shropshire only 78,000, in a total for the United Kingdom of 690,000 tons ; and in 1828 the last of the charcoal-iron furnaces of Sussex (that at Ashburnham) was dismantled.

The next notable advance in colliery engineering was made by T. Y. Hall of Ryton-on-Tyne, who after various unsuccessful attempts at improving the methods of winding, introduced in 1835 the two-decked iron cage travelling between guide-rods and accommodating two iron tubs on wheels. Thus was initiated the modern system of winding ; and the old corf or basket, which had been in use from time immemorial, was rapidly abandoned, with the curious result that the price of hazel-nuts in the London market was at once and permanently lowered.

Explosions, due generally to ' blowers ' locally overpowering the ventilation, were still frequent in the deep and fiery mines of the north, and what was probably the last explosion to blow human bodies from the shaft-bottom to the surface took place in

1817 at the Row Pit (480 feet deep) at Harraton in
the Durham coalfield. Davy-lamps gradually came
into use between 1817 and 1835, though the extended
application of gunpowder in breaking-down the coal-
face rendered their employment no safeguard, as
the gas would fire by the blast of the explosive just
as readily as at a naked candle. Improvements in
ventilation at this time took the form of an increase
in the volume of air. In the north, underground
furnaces were general, but in many of the Midland
pits the fire-lamp was still in vogue, wherever natural
ventilation did not suffice. But in 1835 John Martin
made the fruitful proposal to employ a fan in place
of the furnace, a suggestion that ultimately revolu-
tionized entirely the system of mine-ventilation.

In 1835 a committee of the House of Commons,
after an enquiry into the management of coal-mines,
made certain recommendations for the avoidance of
accidents. While Government inspection and regu-
lative enactments were not considered desirable,
brattices (brick or wooden partitions) in ventilating-
shafts were condemned, and the keeping of maps
and plans was thought worthy of encouragement.
George Stephenson however went so far as to con-
sider that the sinking of two shafts should be made
compulsory.

By the commencement of the Victorian era the
coal and iron industries had fairly entered the

modern period ; the main lines of practice had been already laid down, and subsequent improvements have been for the most part matters of detail. Certain exceptions however call for mention. The risks of explosion at the ventilating-furnace itself had led to numerous proposals for substituting some safer method of producing the air-current ; and in 1828 a Mr Stewart successfully installed a system of steam-jet ventilation at Hendre-forgan near Swansea by discharging a jet of high-pressure steam at the bottom of the upcast shaft ; and for a while a brisk controversy ensued between the upholders of the furnace and the advocates of the steam-jet as the more efficient ventilator. But a rival to both these methods, and one destined to outstrip them, was rapidly coming to the front. In 1837 William Fourness of Leeds brought out an exhausting-fan on the winnowing-fan principle ; and in 1844, by producing a machine capable of exhausting 13,500 cubic feet of air per minute, successfully inaugurated the modern system of ventilation.

A most important improvement in the winding arrangements was rendered available in 1839, when Andrew Smith patented his iron-wire ropes ; and from 1840 onward they came rapidly into use in the northern mines, where previously ropes of hemp, usually flat, and rolled Catherine-wheel fashion, but occasionally round, had been employed in general. The great weight of the hempen ropes had led in some cases to their being woven taper-wise, as at Monk-wearmouth, in 1837, when a new flat rope, 600 yards long, 8½ inches broad at the top, and narrowing to 5½ inches at the bottom, was set up at a cost of £300, with a prospect of its lasting little more than a twelvemonth. In the shallower pits of Shropshire and South Staffordshire, iron chains were usual during the first half of the 19th century; and although their clank and rattle are heard no longer, they may still be seen in the vicinity of the pits, serving the useful purpose of fencing.

About 1850 the double-cylinder engine, without a fly-wheel, was introduced for winding and haulage purposes; and about this period various devices were patented for the prevention of 'over-winding' (*i.e.* pulling the cage up into the framework over the shaft), and for stopping the fall of the cage, should the rope break. Underground, the old and cumbersome practice of conveying several tubs on a wheeled carriage or rolley along the main roads began to be given up about 1841–2 in favour of trains of tubs running on their own wheels along malleable iron edge-rails.

Patent fuels began to receive attention about 1838, the object being to utilize small coal by binding it together with some such material as coal-tar and moulding the mixture into bricks that could be used as fuel. It has long been the practice among the South Welsh country-folk to turn to account the local culm or fine slack by mixing it with clay or lime into a coherent mass capable of being moulded by hand into ' balls '—a fuel certainly more pleasant for kitchen consumption than that manufactured at Westminster about 1819 by one Chabauner, in which a principal ingredient was the sweepings of the streets !

During the latter half of the 19th century many improvements were introduced, some of which will be described in the following pages, such as the modern methods of boring and shaft-sinking, coal-cutting by machinery, and new forms of safety-lamps ; but these are concerned chiefly with the mechanical and engineering departments of coal-mining. The most striking innovation of recent years is the application of electricity in the departments of hauling, pumping, winding, coal-cutting, lighting, signalling, drilling, and shot-firing. As a form of energy it is easily conducted by wire to all parts of the mine, and in this way is much more easily installed than compressed air, which is some-times used as a motive-power. Electricity, however, has the drawback of being a source of some danger from sparking, with the attendant risk of firing the gas, and is not free from the liability of giving fatal shocks to the workmen.

Having now sketched the evolution of mining-processes from the earliest times, we shall next proceed to a description of the operations as practised at the present day, prefacing that description with such geological observations as are necessary for a proper understanding of the various methods of working the coal.

CHAPTER II

VARIETIES, GEOLOGICAL AGE AND ORIGIN OF COAL

SINCE the natural history of coal has been treated of in another volume of this series, a brief summary alone will be attempted here.

Varieties of Coal.—The term coal as used to-day covers a variety of substances differing greatly in physical and chemical properties, in their ages, and modes of formation. But all coal-seams are in the last resort beds of ancient vegetable matter more or less chemically altered, composed chiefly of hydrocarbons (compounds of hydrogen and carbon), and suitable for use as fuel. The common varieties of coal are Lignite, Brown Coal, Cannel Coal, Coking Coal, Gas Coal, House Coal, Steam Coal, and Anthracite. Very closely in the order here adopted they diverge gradually in their chemical, and largely in their physical, properties from Peat, the recent vegetable accumulation of our swamps and moorlands, and approach in character the Graphite of which pencils are made. It must not be inferred from this, however, that all coal originated as peat, and will ultimately be converted into anthracite or graphite as the final result of metamorphic processes (p. 51).

In Lignite the woody constituents are so little altered that their form and structure are usually discernible; while leaves, bark, and other tissues are often well-preserved. Lignite is brown to pitch-black in colour and burns easily, emitting a smoky flame and an unpleasant odour. In the Brown Coals the woody constituents are not obvious to the eye. Cannel Coal is black or brownish in colour, dull and lustreless, clean to the fingers, and can be carved into ornaments. The choir of Lichfield cathedral was formerly paved with squares of cannel obtained probably from a seam in Beaudesert Park near Rugeley. It can be ignited with ease on the application of a burning match, and burns with a smoky yellow flame like that of a candle—hence its name. It is specially valued for making gas, a ton of Wigan cannel yielding over 14,000 cubic feet of gas of 39 candle-power. The House Coals, familiar to us all, are generally more or less lustrous, dirty to the fingers, and tend to split along the bedding-planes, and also to break crossways along joints usually at right angles to each other. Traces of the original vegetable matter are as a rule not easily seen with the naked eye ; but if a piece be broken across it will be found to consist usually of alternating dull and bright layers.

The Steam Coals are nearly devoid of lustre, slow to ignite, evolve little gas or smoke while burning, but give out an intense heat—hence their value for generating steam-power, and their importance from a naval point of view. The best steam coal is obtained from the South Wales coalfield. Anthracite or Stone Coal is, next to graphite, the purest form of natural carbon obtainable—except of course the diamond. It has a brilliant lustre resembling that of graphite, is clean to the touch, and is harder, denser and more brittle than ordinary house coal ; it is difficult to ignite, burns slowly, makes little ash, gives off no smoke, but burns with the blue lambent flame of carbon monoxide. In this country it is obtained almost wholly from the north-western and western districts of the South Wales coalfield and from the southern Irish coalfields. A good anthracite contains as much as 95 per cent. of carbon. It is largely used for hop-drying and malting.

Behaviour during Combustion.—In their mode of burning, coals can be grouped as (1) caking coals,

and (2) dry, free-burning or non-caking coals. Those
of the first group partially fuse and cake together,
and at first emit much flame and smoke, and extrude
bubbles of tarry matter and hissing jets of gas ; but
after these volatile matters have been burnt off,
combustion slackens and in an inadequate draught
comes to a standstill, leaving in the grate a dead
accumulation of unconsumed coke. Their property
of caking, however, gives these coals their value as a
source of coke, as the small coal and slack not suitable
for household purposes can so be utilized. The non-
caking coals burn freely, leaving no coke or cinder,
and evolve no tarry matter. Some house coals, *e.g.*
from the Durham coalfield, are caking ; while others,
as most of those from South Staffordshire, are free-
burning.

Suitability.—The suitability of a coal for any
particular purpose depends chiefly on its behaviour
during combustion, its hardness, and its chemical
composition. A soft coal is wasteful, as in its passage
from the pit to the place of consumption it produces
much small coal and dust, for which little use can be
found. Sulphur (in the form of iron pyrite, ' brasses,'
FeS_2) is usually detrimental ; a pyritous coal during
combustion evolves sulphur dioxide (SO_2), a pungent
gas that not only offends the nostrils but also attacks
metals, such as the bars of grates and furnaces and
the plates of boilers. Lumps of iron pyrite are seldom

allowed to get as far as the coal-scuttle, but the mineral may often be seen as a thin brassy film on the joint-faces of some of the pieces of coal. Pyrite is particularly objectionable in a gas-coal, owing to the vitiating effect of the resultant gas, when burnt, on the atmosphere of the dwelling-room ; moreover, it sometimes contains a dangerous amount of arsenic, which renders pyritiferous coals unsuitable for hop-drying and malting.

Geological Age of Coals.—The geological age of coals is a matter of considerable practical importance. In the British Isles our coals are referable to the Oligocene, the Jurassic, and the Carboniferous systems. The Lignite of Bovey Tracey in Devon is of Oligocene age ; the coals found on the Yorkshire coast, at Brora in Sutherland, and at Kimeridge in Dorset, are all of Jurassic age. But none of these is of much economic value outside its own immediate district ; and, in this country, when we speak of coal, we mean Carboniferous coal, that is, coal found in the Carboniferous systems of rocks, and especially in that division thereof known as the Coal Measures. These consist of a great series (in South Wales as much as 10,000 feet) of alternating conglomerates, sandstones, shales and clays, with ironstones and occasional thin limestones, all characterized by the presence of certain fossils (plants, mollusca, reptiles, insects and fishes) more or less restricted to that series, and giving evidence of conditions that were predominantly estuarine, lacustrine or terrestrial. Within these measures occur all the important coals of England, of Wales and of Ireland; though in the extreme north of Northumberland, in Cumberland, in Scotland and in Antrim, coals are found in the Lower Carboniferous rocks also (*see* Table, p. 47).

The subjoined complete list of the British geological systems, arranged in descending order, shows the positions of those which yield coal.

BRITISH GEOLOGICAL SYSTEMS

Quaternary	{ RECENT (Peat)
	{ Pleistocene
	⎧ Pliocene
	⎪ (Miocene—absent from Britain)
Kainozoic	⎨ OLIGOCENE (Lignite of Bovey Tracey)
	⎩ Eocene
	⎧ Cretaceous
Mesozoic	⎨ JURASSIC (Coals of Brora, Yorkshire Coast, and Kimeridge)
	⎩ Triassic
	⎧ Permian
	⎪ CARBONIFEROUS (all ordinary coals and anthracites)
Palaeozoic	⎨ Devonian
	⎪ Silurian
	⎪ Ordovician
	⎩ Cambrian
Eozoic	Archaean

Wherever the geological succession is complete, the Coal Measures overlie the partly estuarine but

chiefly marine sandstones and shales of the Millstone
Grit, which in its turn succeeds the Carboniferous
Limestone with its rich and wholly marine fauna.
Above the Coal Measures follow the red poorly-
fossiliferous lacustrine and desert-formed rocks of the
Permian and Trias. The position of our coals in
the series of geological systems is thus perfectly
well known, and if coal-seams were present in
the other members of that series—*e.g.* in the Silurian
or in the Trias—there is not the least doubt that, in
a long-settled and surveyed country like ours, they
would have been discovered ages ago. If therefore
we are asked whether coal will be found under a
certain property, the first point to ascertain is : what
system of rocks occupies the surface ?

If the system at the surface is older than the
Carboniferous it follows that as a rule the deeper we
bore or sink, the farther away from the coal shall we
go. Yet such an obvious inference as this needs
emphasis when one finds, as recently as 1912, a
boring for coal being driven for hundreds of feet into
the Silurian rocks of Radnorshire !

In the past, considerable sums of money have
been expended on boring and sinking for coal into
some of the Ordovician and other rocks—both older
and newer than the Carboniferous—on the strength
of their consisting largely of black shales very like
those of the Coal Measures. Even as recently as

ten years ago a level for coal was opened, with some amount of ceremony, in the side of a hill on the outcrop of the Ordovician rocks in Carmarthenshire. No coal had ever been seen to crop out anywhere in the neighbourhood, but it was enough that the beds were sooty black shales. The fact that they were full of Ordovician graptolites, an order of marine fossils that had become extinct long before Carboniferous times, was un-noticed or ignored by the credulous projectors.

If the surface-rocks are newer than the Coal Measures, there are two methods by which it may be ascertained whether Coal Measures lie below. Firstly, a thorough knowledge of general and local geology should be brought to bear on the problem ; but if the data available lead to no definite conclusion, the second method, namely boring, must be adopted. By this means samples of the rocks can be brought to the surface and examined by the geologist (p. 73).

But even where rocks newer than the Coal Measures come to the surface, it does not follow that coal, or even the Coal Measures themselves, will be found below. Firstly, the coal-bearing strata may never have been deposited, as the site may have formed part of a land-area at that time ; secondly, if deposited, the Coal Measures and their contained coals may have been worn off and destroyed prior to the laying down of the strata that now occupy the surface: in both cases a gap will be found in the geological sequence. It may thus happen that a boring, after traversing rocks newer than the Coal Measures, may pass abruptly and quite unexpectedly into pre-Carboniferous rocks.

A consideration of an actual example will make this clear. In view of the approaching exhaustion of the exposed coalfields of Coalbrookdale and South Staffordshire, an attempt was made some ten years ago to ascertain whether coal could be reached at a workable depth beneath the Triassic and Permian rocks that occupy the intervening area, and a boring was put down at Claverley, between Bridgnorth and Stourbridge. Commenced in the upper part of the Permian, and passing through that formation and all three divisions of the barren Upper Coal Measures, it entered the productive Middle Measures at a depth of 1797 feet, with every prospect of success. But after traversing 393 feet of shales, sandstones, fire-clays, ironstone-bands, and several thin and useless coal-seams, all of the usual character, and containing the usual fossil plants, the drill suddenly entered a hard grey rock, which contained *Atrypa reticularis* and other recognizable marine mollusca that demonstrated it to be Silurian. The whole of the valuable lower part of the local Coal Measures, in which the chief coals of Shropshire and South Staffordshire are situated, would thus appear to be absent. Presumably the Silurian rocks, while the coal-seams were accumulating in adjacent districts, had here formed a shoal or a land-tract that was not submerged till towards the Upper Coal Measure period. No survey of the surface could have foreseen such a result ; but it is obvious that in this case not only was the precaution of boring before sinking a very wise one, but also that a knowledge of fossils was of much value in preventing the borers going deeper.

In some cases it has been found on boring or sinking through the overlying newer rocks that although the productive part of the Coal Measures is present, yet the coal-seams themselves have locally thinned-out, or have so deteriorated in thickness or in quality as to be worthless.

If Coal Measures actually occupy the surface, a knowledge of the local geology will generally indicate whether and at what depth coals may be expected. It by no means follows, however, that all Coal Measures contain coal. Though occasional thin seams and streaks of coal are present throughout, the valuable coals are restricted to the lower and middle parts of the Coal Measures, as shown in the following Table, in which the subdivisions of the 'Upper' Measures are those adopted by Dr Gibson and other officers of the Geological Survey as a result of work in North Staffordshire :—

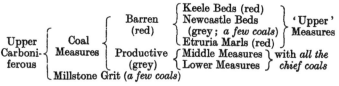

British Carboniferous System, showing Positions of Workable Coals

Upper Carboniferous	Coal Measures	Barren (red)	Keele Beds (red), Newcastle Beds (grey; *a few coals*), Etruria Marls (red)	'Upper' Measures
		Productive (grey)	Middle Measures, Lower Measures	with *all the chief coals*
	Millstone Grit (*a few coals*)			
Lower Carboniferous	Carboniferous Limestone Series (with *coals* in Scotland, North of England, and North of Ireland)			

Organic Remains.—The vegetation of the coal-period as preserved in the coal-seams and their associated rocks was predominantly composed of vascular cryptogams and other flowerless plants of arborescent habit referable to five chief groups, viz. the Lycopodiales, the Equisetales, the Pteridosperms and Filicales, the Sphenophyllales, and the Cordaitales. The first two are represented respectively by our present-day Club-mosses (*Selaginella, Isoetes,* etc.) and Horsetails, and the Filicales by the modern Ferns; but the other groups are extinct. The flora of the period thus presented a very sombre and monotonous aspect; the flowering plants of our modern landscapes, with all their variety of colour, had not yet appeared. Most of the plants grew to a great size, stems of *Lepidodendron* and *Sigillaria* 50 feet in length being not uncommon.

The Lycopodiales were represented chiefly by the

genera *Lepidodendron* and *Sigillaria*. In the first,
the stem is covered with the spirally-arranged
lozenge-shaped leaf-bases. Its 'fruit' was a cone
known as *Lepidostrobus*. In *Sigillaria* the stem is
usually marked by vertical ribs on which the leaf-
scars are placed at intervals one above another ;
sigillarian bark frequently composes the bright bands
in a coal-seam. The root-like organs of both
Lepidodendron and *Sigillaria* are very often found in
the underclays of the seams and are known as *Stigmaria* ; their surfaces are characterized by oval scars
from which rootlets spread into the surrounding mud.

The Equisetales were represented by numerous
species of *Calamites*, whose lofty jointed stems made
up dense thickets along the swamps, as do its nearest
modern relatives the Horsetails along the margins
of our ponds. The stems, which bore narrow leaves
arranged in whorls at the nodes, are generally preserved in the fossil form as casts of the pith-cavity.

The Pteridosperms were plants with fronds
resembling more or less closely those of some recent
ferns, but distinguished by the bearing of seeds and
not merely spores, and by certain anatomical features.

The Sphenophyllales were slender herbaceous
plants resembling the Calamites in habit.

The Cordaitales included large trees characterized
by long strap-like leaves and in habit resembling the
Kauri Pine of New Zealand.

Many of these plants probably formed dense jungles in the low ground bordering the lagoons; but it is likely that on the uplands others flourished, of which few relics have been preserved. Through the moist atmosphere flitted a few primitive mayflies and dragon-flies; while scorpions, spiders and milli-pedes crawled along the rotting stems. There were no birds to prey upon them. In the waters below, a few ganoid fishes disported themselves and afforded sustenance to the salamander-like Labyrinthodonts; while mud-loving bivalve molluscs—*Carbonicola, Anthracomya* and *Naiadites*—closely resembling our pond-mussels, fattened in the slime beneath. Oc-casionally an irruption of salt water brought with it some mollusca of the outer sea, such as species of *Productus, Chonetes, Lingula, Pterinopecten* and *Gastrioceras.*

Use of Fossils.—Not only is a knowledge of fossils of great practical importance to the miner in enabling him to guard against fruitless sinkings in rocks known elsewhere to be devoid of coal, but it is also of use in helping him to distinguish one part of the Coal Measure series from another. The plant-remains form a useful index for this purpose, as Kidston, Arber and Walcot Gibson have shown; for certain plants are restricted to the Upper Coal Measures, while others have not been found above the Middle Measures. The mollusca are still more

C.   4

useful, and can often be depended upon to identify a particular seam in widely-separated coal-pits. So far then from fossils being merely the playthings of the curiosity-hunter, in the hands of the geologist they are of the greatest value in directing the collier to a suitable place for his operations and in protecting him from futile and hopeless undertakings.

Conditions of Deposition.—It is generally agreed that the coal-seams originated from vegetable matter produced by the decay of luxuriant swamps and forests, which spread out from the land into the shallow waters of estuaries, lagoons or lakes, much as do the present-day mangrove-swamps of tropical countries. But two different views have been held to explain the formation of the seam itself. The advocates of the ' growth-in-situ ' theory believe that the seam represents the actual peat-bog or morass itself, and that the underclay generally found below the seam is nothing else than the soil on which the vegetation grew. On the other hand, the adherents of the 'drift' theory believe that the vegetable debris of the swamps was carried out into the lagoon by running water and there deposited like any other sediment. So good a case has been made out by both parties that it appears certain that some coals have been formed in one way and others in another, while it is highly probable that in many cases both modes of formation have shared in the production of a single seam.

Coal Formation.—After a mass of vegetable debris had accumulated, the slow subsidence that affected the region carried the mass below water-level, preserved it from decay, and sealed it up under layers of gravel, sand and mud, where it became converted into a seam of coal. The processes to which this conversion are to be attributed have been discussed at length by Dr E. A. N. Arber. They appear to have been chiefly biochemical, and due to the action of bacteria; they were attended by a loss of oxygen and hydrogen, and an evolution of carbon dioxide (CO_2) and methane (marsh gas, CH_4), the final result being a pulp of hydrocarbons, relatively richer in carbon, in which organic structures are largely obliterated. Whether the resulting coal is sapropelic (such as cannel), or humic (*e.g.* house coal), or anthracitic, seems to have been determined chiefly by the extent to which bacterial action had proceeded before being arrested by the poisonous organic acids to which that action gave rise, though differences in the nature of the vegetation no doubt had much influence. There is good evidence that the conversion of the vegetable debris into coal took place soon after its entombment, and that subsequent heat and pressure, consequent on its burial deep in the earth-crust, did little more than consolidate and harden it.

CHAPTER III

THE COAL MEASURES AND THE COAL-SEAM

Lithology.—The productive Lower and Middle Coal Measures of Britain consist of a great series of conglomerates, grits, sandstones, shales and clays, with bands of ironstone, and numerous seams of coal. These materials were laid down, never far from land, in the shallow waters of swamps, lagoons and estuaries, to which the sea gained only occasional access. As the region slowly subsided, so the accumulating sediments increased in thickness, till in the deeper hollows as much as 10,000 feet had been deposited.

The coarse pebbly materials forming the conglomerates ('pudding-stones' of the miner) are seldom persistent, but are irregularly bedded and lenticular, and show signs of rapid accumulation and repeated sorting by change of currents, with re-deposition not far away. They often form a basement-group to a series of sandstones, which are usually more persistent and can frequently be traced for several miles. Where these coarse materials were deposited uninterruptedly for a lengthened period over wide areas, they constitute an important member of the local Coal Measure sequence, as is the case with the Pennant Sandstone group of South Wales, the Forest of Dene, and Somerset, which attains a thickness of several thousand feet. Few coal-seams occur within these thick sandstone groups. Usually, however, sandstones, shales and clays alternate with each other, and also pass laterally one into the other.

The shales and clays, which are more persistent than the sandstones, and make up the larger proportion of the Coal Measures, are more tranquilly formed deposits of fine-grained clayey matter. The shales ('binds' of the miner) are composed of thin layers, often no thicker than a post-card, the surfaces of which are frequently covered with a film of sand or flakes of mica, which give them a fissile character, so that they readily split into thin slabs, plates, or leaves. The clays ('clunch' and 'clod') are beds of non-laminated mud that crumbles into small irregular fragments. The ironstones are generally nodular concretionary masses of earthy carbonate, in the form of flattened balls ranging up to a foot or more in diameter, or smaller irregular lumps scattered through the shale. In the past they constituted the chief source of our iron-supply, and were worked on a large scale in South Wales, Coalbrookdale, and South Staffordshire.

All these rocks vary in colour from white to intense black, dependent on the amount of carbonaceous matter present. Where exposed at the surface the sandstones, owing to the oxidation of the iron-compounds usually diffused through the stone, tend to weather with rusty-brown, red, or yellow tints ; but the shales and clays lose their grey or black colours less readily. The fossil-remains of plants, abundant in most of the beds, are specially well-preserved in the shales ; while fragments of fern-leaves or shells often form the nucleus around which the ironstone segregated.

The productive Lower and Middle Measures are succeeded in most of the Midland coalfields by an Upper series of relatively barren measures, in which a red colour prevails (see Table, p. 47). In these, coals are rare, thin, and often pyritous ; several limestones occur, not more than a foot or so in thickness, characterized by the presence of annelids and entomostraca (*Spirorbis* and *Carbonia*, etc.). The red Etruria Marls of the Upper Measures are the source of the famous Staffordshire ' blue bricks.'

The coal-seams themselves when viewed on a true scale form a very small proportion of the total thickness of the Coal Measures ; so that in 10,000 feet of strata in South Wales, for instance, the coals in Glamorgan account for only about 124 feet, which, distributed in 48 seams, gives an average of 2 ft. 7 in. for the thickness of each seam. A coal-seam occurs as a definite bed of rock, just like its associated sand-stones and shales, and usually runs for long distances, maintaining its own characters and holding its proper position in the sequence over many square miles.

But the seam is itself usually more or less composite, and consists of bands of coal separated by partings of shale or clay ; while the constituent coal-bands of a single seam may differ among themselves in quality and thickness.

As an example of the kinds of strata passed through in a coal-shaft, the following section of the upper portion of a pit at Polesworth (Warwickshire) may be quoted :

Shaft-section at Polesworth

	Thickness		Depth	
	Ft.	In.	Ft.	In.
Soil		9		
Gravel and sand	16	0	16	9
Blue bind	1	6		
COAL smut	2	0	20	3
Clunch	6	0		
Blue bind	11	0		
Stony bind	7	6		
Strong blue stone	2	3		
Blue bind	10	6		
COAL	1	4	58	10
Stony clunch	6	6		
Stony bind with ironstone balls	9	6		
Clunch and bat with ironstone balls	7	3		
Strong bind	6	6		
Soft bind	1	6		
COAL		3	90	4
Strong bind	10	3		
Soft bind	1	6		
Clunch and bat	3	0		
Stony clunch	6	6		
Blue bind with ironstone	12	10		
COAL : Smithy Coal	2	4	126	9

The ' gravel and sand ' met with to a depth of
16 ft. 9 in. are probably Glacial deposits. ' Bind '
is a miner's term for shale ; ' clunch ' is a tough
clayey rock ; the ' strong blue stone ' is presumably
sandstone but possibly shale ; ' bat ' is a highly-
carbonaceous black shale. The Smithy Coal, being
the only one in the section of any importance, has
alone been dignified by a distinctive title. The coal
at 20 ft. 3 in., being close to the surface, appears to
have weathered to a powdery condition, and is
recorded as a ' smut.'

As an example of a single seam made up of
several coal-bands of different character, the following section of the Barnsley Seam of Yorkshire may
be given :

	Ft.	In.
Coal called the Day Bed	1	0
Parting (fireclay)		2
Coal called the Middle Bed	1	1
Coal called the Low Bed	1	3
Parting (fireclay)		8
Coal and pyrite, called the Clay Seam		7
Coal called Hards	2	8
Coal called Slottings	2	2
	9	7

The coals from the Day Bed, the Low Bed and the
Slottings are gas- and house-coals ; the Hards is used

for coke-making and steam-raising; while the
pyritous coal of the Clay Seam serves for lime- and
brick-burning.

Roof and Floor.—The bed of rock that immediately overlies the coal-seam is known as its
roof. Where it is a sandstone or a hard shale it
affords facilities to the miner in reducing the cost of
timbering the underground roadways and workings.
A soft friable roof will often give such trouble as to
make a good seam of coal unprofitable to work. The
rock-bed immediately below the coal is known as
the coal-seat, thill or floor. It is usually a bed of
grey clay ('underclay'), a foot or so thick; where
composed chiefly of very finely-divided siliceous mud
free from alkalies it constitutes a fireclay, highly
esteemed for the manufacture of firebricks, crucibles,
melting-pots for the glass-maker, and gas-retorts,
as at Stourbridge in Worcestershire, where the best
fireclay is found below the Thick Coal. A very hard
siliceous floor ('gannister') containing 57 to 96 per
cent. of silica is common in the Lower Coal Measures
of Lancashire and Yorkshire, and is used as a bed
for the hearths of iron-furnaces. A soft clay floor
causes much trouble to the miner, as it swells up
under pressure of the overlying strata and tends to
fill up the underground roadways (p. 29). In the
coal-seat are frequently found the roots (*Stigmaria*)
of some of the trees that grew on, or were drifted to,

the site and contributed to the formation of the overlying coal.

Breaks in the Seam.—The continuity of a coal-seam is liable to be interrupted by a number of causes, some of which were active during or just after its formation, while others did not come into operation till a long-subsequent period. It is obvious that where the swamp or lagoon bordered an elevated land-tract, the coal-seam must have come to an end somewhere along a line of shore, just as a swamp does nowadays, though it is not often possible to point to such an original margin of deposition. In South Staffordshire, however, the Thick Coal, when followed southward of Halesowen and Cradley, has been found to become so earthy and impure through admixture of muddy ingredients as to be useless as a fuel ; and there is no doubt that if followed far enough it would be found to end against the older Palaeozoic rocks (Silurian and Cambrian) that formed the coast of the lagoon. The Upper Measures are known to overlap the lower and to extend farther southward. Elsewhere in the same coalfield, as at West Bromwich, local shoals in the bottom of the lagoon have similarly prevented the deposition of one or more of the coal-seams over considerable areas, a contingency quite incapable, unfortunately, of being foreseen from an examination of the surface. In some places a seam of good coal, far from its original margin, may become worthless over an area some acres in extent owing to an intimate admixture of mud or sediment. Again, a parting of clay, shale, or sandstone in a coal may gradually thicken out at the expense of the coal till little of the coal is left ; or the whole seam may thin away to a knife-edge. When a coal-swamp was submerged and covered with a bed of sediment, the old stream-courses that crossed the swamp, and new channels eroded through the vegetable layer, were filled-in with sand, which now forms a ' wash-out ' which, descending from the roof, may cut out the coal more or less completely for some yards in width. Such a wash-out has been described in the Coleford High Delf Seam in the Forest of Dene. Conversely, a sand-bank occasionally projected upward through the vegetable layer, so that now the coal-seam ends off on both sides of a mass of sandstone rising from the floor. All these interruptions are due to immediately contemporaneous causes.

Other breaks in the continuity of a seam are due to the foldings, tiltings and squeezings that the rocks have undergone during subsequent periods of earth-movement. A considerable tract of productive Middle Coal Measures, with valuable seams of coal and ironstone, along the eastern side of the Coalbrookdale Coalfield, must have been elevated by a gentle folding sufficient to bring it within reach of the

waters of the lagoon or estuary, and may have been
raised even above the waters, and subjected to sub-
aerial erosion during Coal Measure time, for large
areas of the productive measures were washed away
before the deposition of the Upper Measures, which
lie across their edges. Much coal has been thus
destroyed by this Symon Fault, as it is called.

Where the beds have been thrown into undulations it is found that a coal-seam locally thickens
abnormally (a ' swelly '), but in a contiguous part
of the undulation suffers a corresponding constric-
tion, amounting in some cases to complete extinction
(a ' nip-out ').   All these causes, as well as the true
' faults ' to be described anon (pp. 65–7), combine to
harass the miner and damage the fortunes of the
mine ; they call for special precautions and delicate
treatment, and under the name of ' abnormal
places ' carry with them special rates of payment.

Igneous Intrusions.—Lastly, the miner has in
some coalfields to reckon with the devastating effects
of igneous intrusions, which have invaded the coal-
seams and reduced some of them to useless dust.
Molten igneous material, at some period subsequent
to the Carboniferous, forced its way up from below
along more or less vertical fissures (' dykes ') or
pipes (' necks ') in the earth-crust, burrowed a road
for itself between the beds, or raised the overlying
strata into a sort of mushroom-shaped bubble or

laccolite. Whether the molten matter ever reached the surface we cannot tell ; certain it is that it found it easy to follow a bed of coal, which it generally baked to a substance resembling coke or soot. The Rowley Hills in South Staffordshire and the Clee Hills in Shropshire are capped with masses of such basalt (quarried for road-stone as Rowley Rag and Dhustone), while several of the contiguous coal-seams have been invaded by sheets of similar material (' green rock '). Igneous rock has damaged considerable areas of coal about Willenhall, Wednesfield and Bloxwich in South Staffordshire.

CHAPTER IV

COALFIELDS, FOLDS AND FAULTS

Coalfields.—Though the greater part of the area now known as the British Isles was originally covered by the Coal Measures, it is possible to point to certain districts over which it is very questionable if those beds were ever deposited. The high lands of North and Central Wales, the Highlands of Scotland, and the heights in the north-west of Ireland, are probably parts of old land-areas that stood up above the swamps and lagoons of the coal-period. During Lower and Middle Coal Measure

times a land-tract certainly crossed the centre of
England, for there only the Upper Coal Measures
were laid down, as we have seen (p. 58). But the
original shore-lines of the lagoons are seldom preserved or visible ; either they have been removed
long ago by denudation, or they lie concealed under
newer rocks. To the miner it is a matter of small
consequence, as the Coal Measures are now found to
occupy some twenty-five detached areas known as
coalfields or coal-basins, the limits of which have
been largely determined by the plications into which
the rocks were thrown by earth-movements during
Lower Permian time.

Folds.—Originally deposited in more or less horizontal sheets, the Carboniferous rocks were gently
folded into arches (anticlines) and troughs (synclines),
or irregular basins, by movements more or less at
right angles to each other. The arches and elevated
parts of the folds were planed down by detritive
agencies (the sea, rain and rivers, etc.), so that not
only the Coal Measures but also great thicknesses of
the underlying rocks were eroded from their crests.
The formation of such disconnected basins from a
once continuous sheet of sediments is shown in
Fig. 4.

In the centre of a basin the coals are more or
less flat or horizontal, and may lie at a great depth ;
but toward the edges they rise at a considerable

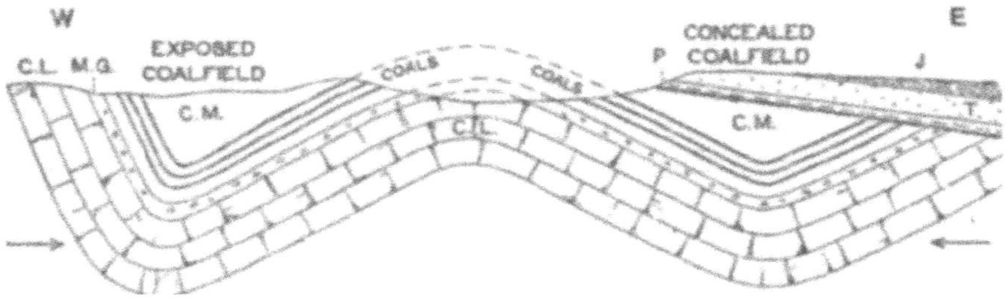

Fig. 4.—Section of a series of Carboniferous strata originally horizontal and continuous, but now folded by compression and forming two detached coalfields, separated by outcrops of Carboniferous Limestone (*C.L.*) and Millstone Grit (*M.G.*).—*C.M.*, Coal Measures; *P*, Permian; *T*, Trias; *J*, Jurassic. The arrows show the direction of the pressure.

It will be observed that the Carboniferous strata now form two synclines or troughs, separated by an intermediate anticline or arch. The oldest Carboniferous beds come to the surface in the core of the anticlines, while the newest are found in the centres of the synclines.

The Carboniferous strata were folded, and then reduced by denudation to a plane, before their edges were covered by the Permian and newer rocks, which have not been folded, but merely tilted gently toward the east.

Steeply-dipping coals, such as those on the western side of the 'exposed coalfield', are sometimes known as 'edge-coals' or 'rearers.'

angle from the horizontal and ultimately reach the surface along what is known as their outcrop. Such a coalfield as that shown on the left side of Fig. 4 is known as an ' exposed ' coalfield, as its margins are visible at the surface. But a coalfield may be wholly or in part concealed by a cover of newer strata laid down unconformably across the eroded edges of the Carboniferous rocks—the folding of the older rocks having been in the main completed before the deposition of the newer cover, as on the right in the figure. The Kent Coalfield is a case in point; the Coal Measures are wholly concealed under 1000–2000 feet of Mesozoic rocks and were discovered solely by deep borings put down at points selected on theoretical grounds by geologists. In the case of the Yorkshire, Derbyshire and Nottinghamshire Coalfield, the western margin comes to the surface in the counties named; but the eastern side is wholly concealed beneath a thick cover of Permian, Triassic and Jurassic rocks, thus resembling Fig. 4; though the Coal Measures have been proved by borings to extend some miles eastward of the exposed area as far at least as the valley of the Trent.

As an exposed coalfield becomes exhausted, the pits and their attendant population and all the unsightly concomitants thereof slowly but surely invade the agricultural borderland, as is already the case in South Staffordshire, Warwickshire, Shropshire, etc., till the coal is worked out, or exceeds the limit of 4000 feet, below which it is probable that it would be impossible or unprofitable to work it.

Overthrust Faults.—Sometimes the compression to which the measures have been subjected has been so intense that the beds have been bent into a vertical or even inverted position, as is nearly the case on the left in Fig. 4. Not infrequently the rocks have

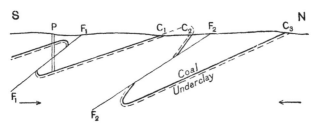

Fig. 5.—Section of a coal-seam affected by compression, which has produced two overthrust faults F_1 and F_2. Each fault dips to the upthrow side. The arrows show the direction of compression.

given way under this treatment, and have slid bodily over one another along an overthrust fault, as in Fig. 5. Here a coal with a southerly dip is cut by two overthrust faults F_1 and F_2; it crops out three times (at C_1, C_2, and C_3), and at C_2 is inverted, the underclay lying on top of the coal. If the surface of the ground had been planed down a little lower, another crop would have been produced near the pit P. In the Somerset Coalfield the Radstock 'slide-fault' has produced a similar effect, and in West Pembrokeshire the measures are so riddled with overthrusts that a single seam will crop out again and again in the space of a few hundred yards. In Fig. 5, a shaft sunk at P would pass through the same seam twice; and there a given area of ground would contain twice the normal quantity of coal.

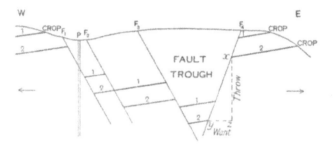

Fig. 6.—Section of two coal-seams (1 and 2) with westerly dip, and affected by extension, which has produced four normal faults (F_1—F_4). Coal No. 1 crops out twice, No. 2 only once. F_1 and F_2 are 'step-faults,' throwing the coals down eastward in two steps. F_2 happens to bring coal No. 1 opposite No. 2; x—z is the 'throw' of F_4; y—z is its 'want' or 'barren ground,' and the angle yxz is the 'hade.' The pit P, being sunk in the 'want,' misses the coals. All the faults dip to the downthrow side. The arrows show the direction of the extension.

Normal Faults.—Usually, however, the beds have suffered much less severely, and form parts of gentle undulations in which extension has been set up. The rocks have snapped along lines of fracture that have allowed the strata to spread out laterally and occupy a greater horizontal extent than at first. Such normal faults are illustrated in Fig. 6, where two coals are shown. It will be noticed that here there is less coal in a given area than would be the case were the ground unfaulted; also that the beds have suffered extension, measured by the sum of the barren grounds or ' wants,' and that no pit would pass through a single coal more than once. The faults all slope or dip toward the downthrow side. The amount of the downthrow may be anything from a fraction of an inch to several thousand feet.

Dip and Strike.—In dealing with an inclined stratum the fundamental conceptions of ' dip ' and ' strike ' must be grasped clearly. The direction of dip is the compass-point toward which an inclined bed exhibits its maximum declination from the horizontal, the amount of the dip being expressed in degrees from the horizontal or, among miners, usually in inches to the yard; a dip of 3 inches a yard means that in a horizontal distance of 36 inches a bed has declined three inches, or at the rate of 1 in 12. ' Strike ' is the compass-direction along which the bed exhibits no dip; strike is always at right angles to the direction of dip. In mining-practice it is called ' level-course,' because as long as an underground roadway in the coal follows the line of strike it remains level. Thus, if a bed dips south or north,

it strikes east and west. A familiar illustration of
these conceptions of dip and strike is furnished by
the ordinary roof of a church. The direction taken
by the drops of water on the sloping tiles is that of
the dip ; the direction of the line of ridge-tiles is the
strike. Dip and strike are quite independent of the
surface-configuration of the country, and hold good
underground. But the course of the outcrop of a
bed across the country depends not only on the
direction and amount of the dip, but also on the
form of the surface. A vertical bed always makes
a straight outcrop, the direction of which coincides
with the strike. An inclined bed cropping out on
a uniform plane, whether that plane is horizontal or
sloping, also makes a straight outcrop. But in all
other cases the outcrop takes a sinuous course dependent on the surface-features.

Joints.—Most rocks are traversed by sets of
fissures that cut through them at right angles with
the bedding-planes and are known as joints. A joint
differs from a fault in that no relative displacement
has taken place along it. Joints usually fall into
two sets, one more or less parallel to the strike, the
other to the dip, and thus between them cut up a
stratum into roughly rectangular blocks. They
afford great assistance to the quarryman and miner,
and generally determine the direction in which
a quarry and the workings in a colliery are to

be laid out. The joints parallel to the strike are usually the more pronounced, and are known as the ' cleat,' ' face,' ' backs ' or ' slyne,' while the joints parallel to the dip are called ' ends ' or ' cutters.' The main roadways in the coal are often driven parallel to the cleat, or ' cleavage ' as it is sometimes called, while the bords or passages from which the coal is taken run parallel to the ' ends ' (Fig. 11, p. 97).

CHAPTER V

PROSPECTING AND BORING

Prospecting.—Enough has been said already (p. 43) to show the futility of searching in Britain for coal in any but the Carboniferous rocks ; and it has been pointed out that the productive measures are confined (except in Scotland, the north of England, and the north of Ireland) to the lower and middle parts of the Coal Measures. In this country geological examinations of limited areas have been carried out in many districts and recorded by private individuals, scientific societies, syndicates or mining prospectors, from the days of George Owen of Hênllys (1602) onward. Some of these investigations preceded, while others have followed, the footsteps of

the State Geological Survey, instituted in 1835. The
results have been embodied in innumerable books,
scientific 'proceedings,' memoirs, and maps on
various scales, so that few districts remain of which
the main geological outlines have not been ascertained ; and it is wholly unlikely that any considerable tracts of Coal Measures arc left to be discovered
on the surface. A geological map of the 'solid'
rocks of the British Isles on the scale of one inch to
the mile has long been completed by the Geological
Survey, and that body is now engaged in a detailed
survey of the coalfields on the six-inch scale, which
permits the tracing of all the rock-outcrops, coal-
crops, faults and folds, as well as the accurate
delineation of the 'superficial' deposits of boulder-
clay, sand, gravel, and alluvia that in many districts
conceal the 'solid' rocks below.

The primary aim of a geological survey is to lay
down on a previously constructed topographical map
a partial or complete delineation of the outcrops
that occupy the surface of a given district, and to
furnish materials for the construction of an ideal
section of the country, such as would be revealed
in the banks of a gigantic trench (a 'longitudinal
section '), or in the sides of a deep shaft (a 'vertical
section '). The degree to which these aims can be
attained depends chiefly on the extent to which the
rocks are exhibited in natural exposures, such as

the craggy sides of mountains, the banks and beds
of streams, the cliffs along the coast, the brows of
inland escarpments; or in artificial sections, such as
wells, mines and quarries, cuttings on railways,
canals and roads, the foundation-trenches of buildings, the surfaces of ploughed fields, and in hedge-
banks and ditches.   It is quite a mistake to suppose
that a geological surveyor has constantly to resort
to digging holes in the ground or sinking trial-shafts
and borings.   On the contrary, all the main features
and most of the details of the geological map of
Britain have been laid down on the evidence of
sections provided by Nature and by ordinary in-
dustrial operations.   But our knowledge of the Coal
Measures would be incomparably smaller than it is
if these rocks had not been explored so thoroughly
by the miner, and the results stored up in plans and
section-books by the mining engineers of the country.
In the future, important accessions to our knowledge
of the underground extensions of the coalfields can
be obtained only by sinking and boring-operations.

The outcrop of a seam of coal is seldom visible,
unless it reach the surface in a perpendicular cliff,
either on the coast or in the rocky sides of a valley
or ravine.   Under the effects of the weather, coal
breaks down into fine slack or 'smut,' which a few
inches of soil or debris will conceal effectually.   A
brook that flows rapidly enough to maintain a clean

rocky bed occasionally affords a glimpse of the coal,
and sometimes the gravel and alluvium along a
watercourse contain lumps of coal that can be
traced upstream to their source, where the seam
itself may be detected.

Ferruginous springs, which throw down a floccu-
lent precipitate of rust-coloured iron oxide, are often
taken to indicate the proximity of coal; but such
a deposit shows nothing more than the presence of
ironstone or iron pyrite (FeS_2) in the rocks below.
On the outcrop of the Coal Measures, ironstone-
bands may or may not be attended by coal-seams;
but in South Wales ferruginous springs are quite
as common on the outcrop of some of the pyritiferous black shales of Ordovician age, from which
coals are wholly lacking.

If the search for the outcrop of a coal be successful, it will be desirable to ascertain its quality
and thickness, the nature of its roof and floor, and
the direction and amount of its dip. If it crop out
with a low dip on steep ground, a tunnel ('heading'
or 'drift') may be driven into it for a few yards so
as to reach the unweathered coal. If on gently-
sloping or flat ground, it will be more convenient
to sink a trial-shaft to the coal on adjacent higher
ground or on the 'dip-side' of the outcrop, i.e. on
that side toward which the coal appears to dip. If
the coal is suspected of having a high dip (over 45°,

say), it may be located by costeaning, *i.e.* by sinking two shafts in a line at right angles to the crop, one on each side of its supposed position, and connecting their bottoms by an underground drift.

If sufficient information has been obtained at the outcrop, and it is confidently anticipated that the coal underlies the property, shafts may be sunk forthwith on a site selected with special reference to underground and surface facilities.

Boring.—If, however, no satisfactory knowledge has been gleaned by the prospecting, and in districts bounded by faults or remote from the outcrops, or where the Coal Measures are concealed under a cover of newer rocks, resort must be had to boring. The operation consists in boring a vertical hole, of small diameter, in the earth-crust, in order to bring up samples of the underlying rocks, coals, etc. The experiment, if successful, should yield data as to the depth, thickness, character and number of the coals. At least three boreholes are necessary, however, to ascertain the strike and dip of the beds, and they should be placed in the form of a triangle, and at sufficient distances to test the whole of the property concerned.

Borings are conducted on two different principles: percussion, and rotation. On the first, different forms of chisels are employed, fixed to the ends of solid rods. By raising the chisel a few inches,

and allowing it to drop on to the rock, a hole is
gradually chipped out. A slight turn is given to
the rods before each fall, so that the hole is kept
circular. As the hole deepens, more rods are screwed
on at the top. In soft strata the hole must be lined
with iron or steel tubes, to prevent the sides falling
in. The powder and fragments of rock produced
by the percussion are brought up at frequent intervals by a cylindrical tool called a sludger ; and as
the borehole usually fills with water oozing from the
porous beds, the samples of stone come up in the
form of mud, in which a sharp look-out must be
kept for fragments of coal. By carefully noting the
length of rod required to pierce each fresh stratum,
a record or ' section ' of the boring is procured. The
defect of the method lies in its bringing up mere
mud and chippings, which afford little information
as to the dip, the lithological characters, or the fossil-
contents of the rocks.

For these reasons the rotatory methods are always preferable. In these the rods are hollow, to
allow of a supply of water being conducted to the
cutting-tool, which consists essentially of a hollow
cylinder, the lower edge of which is armed with hard
minerals (usually rough impure diamonds), or with
teeth like those of a saw. In another method,
chilled steel shot are put down the hole and, find-
ing their way beneath the cylindrical cutter, act as

a rasp—much as primitive man employed sand, water, and a hollow stick to bore holes in his stone axe-heads. By giving a continuous motion (at 200 or 300 rotations per minute, by steam-power) to the cutting-tool, an annular groove is cut in the rock, with the formation of a solid core, which is embraced by the cylinder as the cutting edge descends. By raising the apparatus, the core—in pieces several feet long—can be brought to the surface for examination. The diameter selected for the initial part of the boring will depend on the depth to be attained, and may be as much as 26 inches. As the boring proceeds, the diameter is reduced in several stages, determined by the length of boring that requires to be lined ; so that the lowest cores raised may have a width of only one inch or so, though cores less than four inches in diameter are of little real value. A liberal supply of water conducted down the hollow rods and escaping under the cutter rises to the surface again and so flushes the sediment out of the boring. Unfortunately in soft beds, such as clays and coals, it frequently does this so effectually that no core is left, and proof of the character of this part of the section is to be obtained only by carefully collecting the washings brought to the surface in the escape-water ; and unless great care is exercised, a coal several feet thick may be wholly overlooked.

In Germany borings have been put down to depths of over 6000 feet, the operation lasting several years. A recent boring at Heswall, on the west coast of Cheshire, was carried down to 3362 feet in Trias and Coal Measures in 10 months by Brejcha's method. The cost of percussive boring in Coal Measures is usually quoted at 7s. 6d. per fathom for the first five fathoms ; 15s. a fathom for the second five fathoms, and so on. The Diamond Rock-Boring Co.'s price is 8s. per foot for the first 100 feet ; 16s. per foot for the second 100 ; 24s. for the third 100, and so on.

It would be supposed that cores costing so much money and trouble to obtain would be at once labelled with their depths, and laid out in proper order under cover, so as to protect them from the weather and permit of their thorough examination by a geological expert. But too often everything but the coal itself has been treated with scant courtesy as being of no ' practical ' importance—a somewhat short-sighted policy in face of the fact that the coal itself often yields no proper core, and that a reasonable probability of its presence in the boring may depend on some peculiarity in the associated strata appreciable to the geologist alone. And when it is borne in mind that the journal of the boring is entered up in some cases in quite misleading or unintelligible local terms, and that fundamental

differences between certain of the beds are apt to be
overlooked by the man in charge, it is clear that the
preservation of every part of the core is a matter
of prime importance, even from the view-point of
the colliery projector, not to mention that of the
scientific investigator. Yet so far is this matter
neglected that, even where plenty of space is available, the cores are sometimes piled up one above
another out in the open, where the first lengths are
soon buried out of sight, and the whole pile converted by frost and rain and the trampling of cattle
to a heap of useless debris even before the boring
is finished, and reported on by the expert.

CHAPTER VI

WINNING THE COAL

Winning the Coal.—Satisfactory evidence having been obtained that coal underlies the property,
a decision must be made as to whether it is to be
won by shafts, levels, slants or drifts. If the ground
is comparatively flat, and the measures have only a
slight inclination, as is the case with many of our
coalfields, or if the area to be worked is bounded by
faults or concealed under a cover of newer rocks
so that the coals nowhere crop out, shafts will be

adopted. If, however, the country is deeply trenched
by valleys (as in South Wales and the Forest of
Dene), on the sides of which the coals crop out with
little or no dip, it may be advantageous to retain
the ancient method of driving day-levels from the
outcrop. If the coal rises steeply to the outcrop
on the side of a valley or along the foot of an escarp-
ment, as on the northern margin of the South Wales
Coalfield in Carmarthenshire, where the coal to be
worked underlies a great thickness of barren measures, it is usual to avoid the unremunerative outlay
on shaft-sinking, and to win the seam by ' slants,'
'slopes' or 'slips,' *i.e.* inclined tunnels following the
coal downward from its outcrop. In special circumstances, where high-dipping beds crop out in
hilly ground, the coals may be reached by cross-
measure drifts cut in the rock and ascending or
descending in the measures as the case demands,
much as metalliferous veins are won. These different methods of winning the coal, and some combinations of them, are shown in Fig. 7.

Shafts and Sinking.—Where the measures are
known to have a small dip, and shafts are adopted,
their position will depend largely on the facilities
afforded on the surface for the construction of railways, canals and roads, or on the proximity of a
navigable river, so that the coal can readily be sent
away to its destination. For the land-sale (to supply

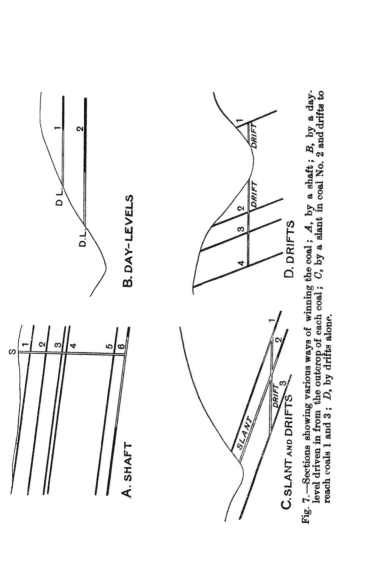

Fig. 7.—Sections showing various ways of winning the coal; *A*, by a shaft; *B*, by a day-level driven in from the outcrop of each coal; *C*, by a slant in coal No. 2 and drifts to reach coals 1 and 3; *D*, by drifts alone.

local demand) a good road is essential.   Where the
measures have a moderate dip, the winding and
pumping shafts are placed 'to the dip' or deep, *i.e.* in
that part of the property toward which the coals dip,
so that gravity may be turned to account in carrying
coal and draining water to the bottom of the shaft.
A ventilating shaft may with advantage be placed ' to
the rise,' as the air-current may be thus assisted in
its ascent.   There must be at least two shafts, placed
not less than 15 yards apart.   In the old days, when
the coals to be worked lay at a small depth, a colliery
would sink half-a-dozen or more shafts at a few
hundred yards apart, as in South Staffordshire ; but
now that the coals have to be reached at much
greater depths, a pair of shafts is made to do duty
for a much larger area, to the no small advantage of
the surface-appearance of the country.

Shafts are rarely less than 10 and sometimes as
much as 20 feet in diameter.   Usually they are circular, as presenting the greatest resistance to lateral
pressure, and as being easiest to sink and to line
with brickwork or cast-iron plates.

The appliances for sinking consist of various
tools, such as picks, shovels, hammers, chisels,
blasting-apparatus, etc., buckets or hoppers for
conveying men and materials up and down, and
some form of winding-gear.   The water encountered
may be too much for the buckets, and may need to

be pumped. For winding, a small temporary engine is installed, from which a steel rope passes over a pulley in the head-gear and is connected with the bucket by a hook.

In ordinary circumstances, as when the Coal Measures lie immediately below the surface, the procedure is as follows. After a depth of 6 feet or so is attained, the shaft is lined with strong and carefully constructed timbering, to prevent the soft soil and subsoil or loose superficial gravels from slipping inward on to the sinkers. Another 6 feet or so are now excavated, and more timbering put in below the first lot, and so on till the first strong bed of stone—the ' stone-head,' as it is called—is reached. On this a curb of wood or of cast-iron is placed and tightly wedged against the shaft sides, so as to form the first course of a brick wall, which is gradually built upward to the surface, and the timbering removed. To build the wall, the masons stand on a circular platform suspended in the shaft and capable of being raised as the work progresses. Sinking is then recommenced. For the first few feet the sides of the shaft are kept flush with the inner face of the overlying wall, but are then cut back to the full width, so as to leave the first section of masonry supported on a bracket or shelf of rock. At a convenient depth, a fresh section of walling is begun, and carried upward to the rock-bracket, which is then carefully removed till the two sections of walling meet end to end.

Sometimes, however, quicksands, usually water-bearing, may form a thick cover over the Coal Measures, and may need special treatment; for, generally speaking, the softer the beds, the more troublesome they are to sink through. One method consists in constructing a strong water-tight open cylinder of wood, of the diameter of the shaft, and armed with a sharp iron edge. This cylinder is then set upright in the sand, and allowed to sink while the sand within is excavated down to the stone-head, upon which the walling can be commenced. A modification of this method is to employ a cast-iron cylinder built up of segments bolted together at flanges on the inner sides. This cylinder is then left as a permanent water-tight lining to the shaft. Other methods depend on ingenious applications of freezing-mixtures to the water-logged sand. In one (Poetsch's method), the sand underlying the site of the sinking is frozen into a solid mass, which can then be sunk through in the ordinary way. To do this, a number of water-tight closed wrought-iron tubes are forced down through the sand till they reach the stone-head. Within each tube a narrower inner tube with openings at the bottom is let down to the same depth. The upper ends of the inner tubes are then connected with a refrigerator and force-pump, by which a freezing liquid is forced downward to the bottom in a continuous current. The liquid escaping at the bottom of the inner tube returns by the outer one to the refrigerator. By these means a column of frozen sand grows round each tube till the whole mass is rendered solid. By a variation of this method the tubes are arranged in a ring round the site of the shaft (Gebhardt and Koenig's method), and a wall of frozen sand produced, within which the unfrozen sand can be excavated down to the stone-head.

In sinking through hard beds, blasting must be resorted to. Shot-holes of an inch or two in diameter are drilled in the stone to a depth of four or five feet, and are arranged in two rings, one round the centre, the other near the edge of the floor. The holes in the inner ring are drilled obliquely, so as to approach each other in order to blow out an inverted cone of rock. The outer ring of holes blows the surrounding mass inward. The holes may be drilled by hand in the manner usual in stone-quarries, where one man holds a long chisel in position, while one or two companions strike it. After each blow the chisel is rotated through a small angle. More rapid progress is obtained by using drills actuated with compressed air or electricity. The holes are then cleaned and charged with an explosive cartridge, to which is attached a length of slow-burning fuse. An electric arrangement however has the advantage of allowing a greater number of shots to be fired simultaneously, while there is no ' hanging fire ' and seldom any ' miss-fire '; moreover, the shots can be fired from any distance.

One of the most serious difficulties that presents itself to the sinker arises when a bed of water-logged sandstone or sand is encountered at some depth in the shaft. In sinking through the Triassic and Permian rocks on the eastern side of the great Northern and Yorkshire coalfields, enormous trouble has been caused by a thin bed of quicksand at the base of the Permian rocks—3000 gallons of water per minute having been encountered in the Monk-wearmouth shafts. Such an amount of water if allowed to descend the shaft would impose a very grievous burden on the pumping-plant, so must be held back by means of tubbing. Nowadays tubbing takes the form of heavy cast-iron plates, each of which is strengthened with ribs and brackets on the side facing the rock, provided with flanges to facilitate fitting, and pierced with a central hole. The first ring of tubbing-plates is laid on a wooden foundation placed on some strong bed of rock below the watery stratum ; the tubbing is then built up ring by ring as far as a similar bed above the watery zone, and finished with a wooden curb wedged tight against the rock above. All the joints between the

iron plates are then wedged up tight, beginning at
the bottom. The hole in each plate, at first left
open to allow any imprisoned air to escape, is then
plugged with wood, and the space behind the tubbing filled in with concrete. A vent-pipe is sometimes inserted in the upper ring of plates and carried
some way up the shaft, to avoid dangerous air-
pressure. In this way not only may a heavy feeder
of water be kept back, but the drying-up of surface-
wells, streams and springs is avoided, to the no
small advantage of the inhabitants.

Sometimes in dealing with hard beds that
yield more water than can be kept under by the
pump, the Kind-Chaudron system is adopted. The
shaft is bored out, first of small diameter, then to
the full size, on the percussion principle (p. 73)
with a heavy circular cutting-tool or trepan, oper-
ated from the surface, during which process no
pumping is done. As the boring advances, cast-
iron permanent tubbing is lowered down the shaft.
The bottom ring of the tubbing is fitted with a
sliding case packed with moss or oakum, which is
so compressed between the tubbing and the rock
below that a water-tight junction is secured. The
water is then pumped out, and the space behind
the tubbing filled in with cement.

Some of the deepest shafts in this country are
those of the Florence Colliery at Longton in North

Staffordshire, which reach the Yard Coal at the
enormous depth of 2490 feet (830 yards) ; while the
Ashton Moss shafts near Manchester are 2880 feet,
or over half-a-mile, deep.

It is usual to carry the shaft a few yards below
the lowest coal to be worked, so as to afford stand-
age for the water. This extension is known as the
sump ; and from it the pumps lift the water to the
surface. The seam being reached, much remains to
be done before coal-getting can be commenced. In
order to afford a firm foundation for the shafts, and
so to avoid collapse and damage to the surface-plant,
a considerable area of coal known as the shaft-
pillar, the size of which will depend on the depth
of the seam, the goodness of the roof and floor,
and the hardness of the coal, must be left unworked
around each shaft in every seam. In a seam 300
yards in depth the shaft-pillar should have a diameter of at least 90 yards. It is usual to place
the two shafts withinrsay 100 yards of each other,
so as to allow of the surface-plant (engine-houses,
offices, etc.) being concentrated ; but they must
not be less than 15 yards apart. The one by which
the ventilating air-current ascends is called the upcast shaft ; the other, by which fresh air descends,
is the downcast, and is usually the one by which
the winding is done. As soon as the shafts reach
the seam to be worked, a communicating passage

must be cut in the coal from one to the other, so as to establish the ventilating current.

Driving Levels.—The next points for decision are the system on which the coal is to be worked, and the direction to be given to the main roads or levels, by which the shafts will communicate with the most

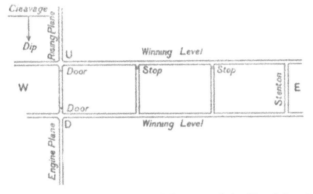

Fig. 8.—Plan showing upcast and downcast shafts *U* and *D*, pairs of winning levels advancing eastward and westward, a rising plane going north, and a dipping plane (engine-plane) going south.

distant parts of the mine, and by which men and boys, horses, trams, air and materials will constantly go to and fro or, as the miner terms it, ' inbye ' (from the shafts into the workings) and ' outbye ' (out of the workings and toward the shafts). Levels are usually driven out from the shafts on both sides,

parallel to each other, in pairs or triplets, with a rib
of coal 20 yards or so wide between them. As they
advance, cross-headings (stentons) are cut from one
to the other every 30 or 40 yards; and as a fresh
stenton is cut, the previous one is closed with
a stopping, so as always to maintain an air-course
along the whole length of the levels (Fig. 8, p. 87).

In opening out these main roads, ample room
must be secured for a hundred yards or so from the
winding-shaft for the construction of the necessary
sidings and railways, along which the full tubs will
be brought up on their way to the surface, and
empties sent off into the mine. Usually this most
important part of the workings is arched with brickwork, and plenty of headroom obtained by ripping
down some of the roof. The main roads or levels
are, as their name implies, level roadways cut in the
coal; and from what has been said already (p. 67)
they must, in an inclined seam, take the direction
of the strike, and cross the direction of dip at right
angles. Also, they usually coincide with the cleat
or main set of joints in the coal. They are not,
however, made perfectly level, but are given a slight
inclination of about 1 in 130, so as not only to cause
the water to flow back to the shaft, but also to assist
the loaded trams in their journey thither. They
are usually driven 7 to 10 feet wide and 6 or 7 feet
high.

Where the seam has a considerable inclination, inclined planes must be driven in the coal, at right angles to the main levels, one toward the rise, and another toward the dip—the latter being known as the engine-plane, since the coal from the dip-workings will be pulled up this plane by some system of mechanical haulage.

If the proprietors of the colliery are prepared to waive any immediate return on the capital outlay, and if other circumstances (pp. 105–6) are favourable, the levels and inclined planes may be driven right away to the boundaries of the property, and the coal then worked back toward the shafts by the Longwall Retreating method (p. 105), leaving the empty space ('goaf') behind. The usual procedure, however, is to adopt a method by which coal can be got as soon as the levels have advanced beyond the shaft-pillar. Beyond the shaft-sidings, the main roads are not so roomy and are not usually arched with brickwork; but as it is essential that they should be kept free from obstruction, their roof and sides are supported with stout timbers, unless the roof should be strong enough to render this unnecessary. A soft shale roof at a great depth, and a soft coal in the sides, require much timber and need constant attention and repair. A very usual arrangement is to place at frequent intervals two upright posts, usually of pine, fir or larch, and

about 6 inches in diameter, one on each side of the
level, and slightly inclined toward each other at the
top, with a thicker crown-piece or lintel laid across
them.   Much pit-wood is imported from Norway and
Sweden.   Owing to the moisture and warmth of the
air underground, the timber is subject to rapid decay,
to prevent which various chemical treatments such
as creosoting have been introduced.

Driving through Faults.—After a level has been
driven in the coal for some distance, it may en-
counter a fault (pp. 65–7), by which the coal is
thrown out of sight, and has to be sought for ; and
it is important to grasp the principles on which this
search is based.   The fault may be a clean-cut fracture with no appreciable space between the end of
the coal and the beds beyond ; or it may be marked
by a variable thickness of ' fault-rock,' *i.e.* layers
of shattered and powdered rock and coal-dust (the
' leader ' of the fault), jammed between the two
cheeks of the fracture.   It has been explained already (p. 67) that in normal faulting, such as is
usually met with in our coalfields, the fault dips
toward the downthrow side.   If, therefore, the pitman in driving a level meets with a fault that dips
away from him, and forms an obtuse angle with the
roof of the level, he assumes that on the other side
of the fault the coal has been thrown down.   He
may now attempt to reach the coal by one of several

methods. If the beds are approximately horizontal, he must give his excavation through the measures

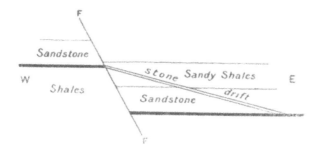

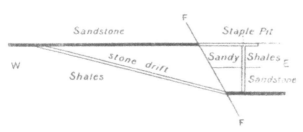

Fig. 9.—Sections showing methods of recovering the coal beyond a downthrow fault *F*, in horizontal or low-dipping beds. The drivage in the coal is proceeding eastward.

beyond the fault a regular downhill gradient suitable for haulage, and continue it till it cuts the coal on the downthrow side, as in the upper section in Fig. 9.

Such an excavation is called a stone drift, as it is
cut through stone and not coal.

Instead, however, of driving blindly downhill to

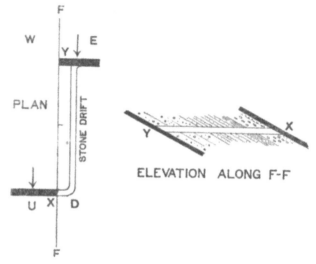

Fig. 10.—Plan and elevation of underground workings to show how
an inclined coal thrown down by a fault *F* at *X* may be recovered
without departing from a level course.   The arrows show the
direction of dip.   The elevation, taken along the fault, shows that
a horizontal cross-measure or stone drift from the coal on the
upthrow side at *X* will reach the coal on the downthrow side at *Y*
U, upthrow side of fault ; *D*, downthrow side.

reach the coal, as in the case just described, he would
do better to carry the level forward a few yards
beyond the fault, and then put down a borehole, or

sink a 'staple-pit' (as in the lower section in Fig. 9), till the coal was reached. He would then go back along his level and cut a stone drift with a downhill gradient requisite to reach the end of the coal on the downthrow side.

Supposing, however, that the measures have a considerable dip, the miner may reach the coal without abandoning his level-course—an important consideration if the level is utilized to carry water back to the shaft. Assume, for instance, that the level is advancing eastward (Fig. 10, p. 92) in beds that have a southerly dip, and a downthrow fault is passed through ; the pitman knows that the coal will be under his feet and will rise to the north of him. By turning his level northward, toward the rise, and cutting a horizontal stone drift descending in the measures (a 'descending drift'), he will ultimately reach the coal.

Again : it may be necessary to make a communication between one seam and another, quite independently of any shaft. This is done by cutting a stone drift, which may be rising, dipping, or horizontal (Fig. 7 C, p. 79). Driving a stone drift usually requires the assistance of blasting operations, which are conducted in much the same manner as in shaft-sinking (p. 83).

Old Workings.—In driving out levels or working-places toward old workings, great care must be

exercised lest the water or gas, with which they may
be charged, be suddenly tapped and let into the new
workings.   On approaching old mines, horizontal
boreholes are driven forward in the coal, with others
going off at angles ; and the exploring heading must
not have less than 5 yards of straight-on boring
ahead of it.   If a boring taps water or gas, it must
be at once closed with strong wooden plugs, and in
that direction no further driving must be attempted.
Lack of information as to the situation of old workings is often a cause of much anxiety to the colliery
officials.   If much water makes its way into the
workings from abandoned mines, it may be necessary
to dam it off.   If the water is under no great ' head,'
stout wooden battens like railway-sleepers, laid one
above another so as to make a wall, are inserted in
grooves cut into the two sides of the heading.   At
a foot or so in advance of the first, a second wooden
wall is erected.   The space between the two is then
tightly packed with clay, which will form a water-
tight barrier.   If, however, the water to be ex-
cluded exerts a great pressure, a wooden or a brickwork dam is constructed in the form of an arch laid
on its face, with the convex curve of the arch toward
the water.

CHAPTER VII

WORKING THE COAL

WHEN the shafts have reached the coal intended
to be worked, and the winding-engines and ventilating arrangements have been installed, the shaft-
sidings and shaft-pillar laid out, and the main levels
and inclines driven forward, the colliery will be in a
position to begin working the coal ; and a method
of work must be decided upon that will produce the
largest amount of marketable fuel at the least cost
and with the greatest safety to the men.   The choice
will depend on the character and thickness of the
seam, its depth and inclination, and the nature of
its roof and floor.

Methods of Working.—The usual methods, although capable of numerous modifications and
combinations to suit special conditions, are :

(1) *Bord-and-Pillar*, otherwise called Bord-and-
Wall, Pillar-and-Stall, Post-and-Stall, Stoop-and-
Room.

(2) *Longwall.*

A description of several other less usual methods,
such as the Single-road Stall and the Double-road
Stall systems of South Wales, the Wicket system of
North Wales, and the Hill system of Warwickshire,
would carry us beyond the purview of the present

volume ; but a brief account of the Square-work of
South Staffordshire will be given.

In Bord-and-Pillar working (Fig. 12, p. 99),
which is doubtless the earliest and most obvious
method, and is the usual one in Northumberland
and Durham, the procedure consists, first, in removing the coal from two sets of working-places,
called bords and headways, driven at right angles
to each other, and forming between them rectangular
pillars of coal sufficiently large to support the roof.
This operation is known as ' whole-working,' i.e.
working in the whole or unbroached seam ; and in
earlier days the pillars so formed were abandoned
in the mine as soon as the boundaries of the property
were reached, as it was found impossible to remove
them with safety.   In modern practice, however, a
second operation, known as ' broken ' or pillar-
working, is performed, by which the pillars themselves, purposely left large at the whole-working,
are removed more or less completely.

In Longwall working (Fig. 13, p. 104), prevalent in Yorkshire, Derbyshire and the Midlands, the
whole of the coal is removed at one operation along
a continuous face or ' wall,' the overlying strata
being allowed to settle down in the vacuity (' goaf '
or ' gob ') behind. This system may be worked
either as (a) ' Longwall Advancing,' i.e. working
away from the shafts and toward the boundaries,

or (*b*) ' Longwall Retreating,' *i.e.* working back from
the boundaries toward the shafts.

Bord-and-Pillar.—In Bord-and-Pillar working,
the ' whole ' working is carried out by driving one

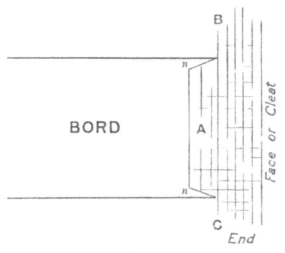

Fig. 11.—Plan of a bord advancing on the ' face ' of the coal. The
   block of coal *A* having been undercut, and ' nicked ' at the sides
   (*n, n*) as far as the ' face ' *B—C*, can readily be removed.

set of wide excavations called bords, which yield the
bulk of the coal, and another narrower set at right
angles called headways, which are used mainly for
ventilation. The two sets between them form
rectangular pillars of coal, to be removed by the

o.   7

' broken ' working later on.   The bords (4 to
7 yards wide) are usually driven wider than the
headways, and are generally cut across the cleat or
' on the face ' ; the headways (2 to 4 yards wide)
are parallel to the cleat or ' on the end.'   The
reason for this is that in cutting the bords, which
as compared with the headways yield the bulk of
the coal obtained by the ' whole ' working, the coal
is much more easily brought down by the hewer if
the main joints cross the line of his excavation.
Thus in Fig. 11, the joint, cleat, or face B–C will have
sliced off the coal already, and all the hewer has to
do is to cut a horizontal groove between the coal
and its floor, and a vertical ' nick ' n, n, at one or
both sides, when the block of coal A will be ready
to be pulled out.   If, however, the bord were advancing on the ' end,' there would be no strong
joint at the back of the block, nor could the hewer
get behind the coal to ' nick ' it.

It will be gathered that a small proportion only
of the coal is obtained by the ' whole ' working
(5 to 30 per cent.) : the bulk of it is left in the pillars.

The manner in which the coal is hewed is as
follows. The hewer with his pick first undercuts
('holes' or 'kirves') the seam across the full width
of the working-place (bord or headway, as the case
may be). This horizontal groove will be about a
foot wide at the face, but will taper inward to a

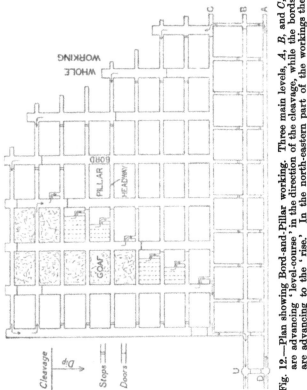

Fig. 12.—Plan showing Bord-and-Pillar working. Three main levels, *A*, *B*, and *C*, are advancing 'level-course', in the direction of the cleavage, while the bords are advancing to the 'rise.' In the north-eastern part of the workings the 'whole' coal is being worked; in the north-western part, the pillars are being worked off in bordway lifts, and the roof is settling down into the goaf. The arrows show the direction of the air-current. *U*, *D*, upcast and downcast shafts. (After C. Pamely.)

knife-edge, and is cut as much as 3 feet under the
seam.   The holing may be done in the coal, or in some
soft parting, or in the underclay.   In holing below
the coal, the hewer lies on his side and swings his
pick horizontally (in a temperature sometimes as high
as 80° F.!); and where the holing is deep and the coal
tender, it may be necessary to prop up the coal with
sprags to prevent it falling forward on to the hewer.
In driving narrow headways and levels, the coal may
need to be nicked, after the holing, with a vertical
groove on one or both sides, before it can be got
down with the pick, with various kinds of wedges,
or by the use of explosives ; the aim of the hewer
being to get the coal in as large (' round ') pieces as
possible.

Explosives are classed according as they are
capable of ignition by heat or by detonation.   Gunpowder is a familiar example of the first class ; it
explodes only on being heated, and is usually fired
with a fuse.   Dynamite and gun-cotton are examples
of the second class, and need a shock to explode them,
although they burn quietly on the simple application
of heat ; they are fired by being placed in contact
with a small charge of some other high explosive
(the detonator or cap) capable of being ignited by
electricity.   For use in coal-blasting, an explosive
(1) should be safe when handled or carried about,
(2) should give off as small and as brief a flame as

possible, and so be not liable to ignite fire-damp
or coal-dust, (3) should evolve no serious volume of
inflammable or poisonous gases on explosion, and
(4) should eject no incandescent sparks. Common
gunpowder offends against three of these canons of
safety, and for that reason a great number of explosives have been introduced that claim to be more or
less flameless, or to evolve no combustible or poisonous
gases.

The amount of timbering required in a working-
place will depend on the nature of the roof and the
width of the bords and headways. Upright props
are used, with a cap at the top, or chocks, consisting
of piles of horizontal posts placed crosswise two and
two. As the working-face advances, the chocks and
props at the edge of the goaf are withdrawn and used
again if sufficiently sound. The bord is ventilated
by dividing it into two parts by a vertical air-tight
partition of brattice-cloth, canvas, or wood, and
conducting the air along one side of the brattice,
round its end, up to the working-face, and back into
the headway (Fig. 15, p. 115).

The size of the pillars formed by the whole-
working is governed by the depth of the seam below
the surface ; for as the thickness of the overlying
strata increases, the pillars must be left larger to
prevent their being 'crushed' to useless slack ;
further, if the coal is soft, the pillars need to be

larger than if the coal were harder ; and if the roof
and floor are soft, the pillars must be large enough
to prevent ' creep,' *i.e.* bulging-up of the floor of the
excavations (p. 29).   It is customary nowadays to
leave pillars 20 to 50 yards long and 10 to 40 yards
wide.

The second operation in the bord-and-pillar
method is the ' broken ' working, or removal of the
pillars.   This may be deferred till the whole-working
has reached the boundaries, but is usually commenced soon after the whole-working has attained
a safe distance from the shafts, and follows up the
advancing whole-work at a convenient distance, if
possible before the roofs of the bords and headways
have fallen in, the empty space or goaf left by the
pillar-working being utilized for the stowage of
rubbish and allowed to fill up by the settling-down
of the roof.

There are innumerable methods of removing a
pillar.   The system generally pursued is that of
taking off bordway slices ('lifts') driven halfway
along the pillar from each headway ; or in the case
of long pillars, by driving a narrow heading across
the middle of the pillar, and then carrying lifts right
and left from this as well as from the headways.   In
working the pillars a regular line of advance (generally making 45° with the headways) should be
maintained between the goaf on the one hand and

the unworked pillars on the other, special care being
taken to avoid leaving a pillar or a stump of coal
behind in the goaf, as its removal would then be
attended by much risk and difficulty. Pillar-
working requires the roof to be specially well tim-
bered.

Longwall.—In Longwall Advancing (see upper
part of Fig. 13, p. 104), as soon as the boundaries
of the shaft-pillar have been left behind, the removal
of the coal along a continuous face or ' wall ' can
be begun. The face may be more or less straight,
or stepped, according to the nature of the cleat, and
sometimes extends for a mile in length. The coal is
undercut as in bord-and-pillar work, but machine
cutters are sometimes employed. The lower edge of
the undercut coal is held up by sprags, and the roof
is upheld by chocks. As the face advances, the
timbering is moved forward with it, whereupon the
roof behind falls, and fills up the goaf. Communication with the shafts is kept up by maintaining
stone-walled passages (gateroads), at least 6 feet
wide, from the face through the goaf, the necessary
stone being obtained from the roof or floor. The
walls of the passages are called packwalls, and should
be well and solidly built, and at least 6 feet thick.
As the gateroads become longer and the expense
of keeping them in order becomes serious, one or
two chief roads are maintained, from which cross

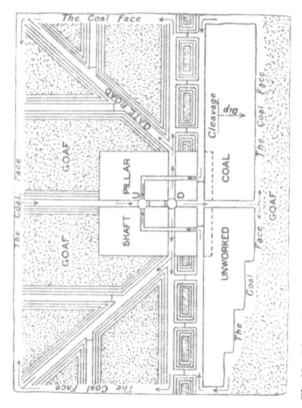

Fig. 13.—Plan of Longwall working, advancing northward to the rise and retreating northward from the dip. The gateroads through the goaf are lined with packwalls.

gateroads are constructed up to the face, and the old roads abandoned. A well-stowed goaf assists materially in the maintenance of the gateroads, and diminishes the timbering required.

In Longwall Retreating (see lower part of Fig. 13, p. 104), a method of working that may be adopted where an immediate yield of coal is not essential, the levels, airways, roads, etc., are driven out from the shafts to the boundaries, whence the coal is worked back toward the shafts by a longwall face. The roof settles down behind. The gateroads are thus all in the unworked coal, so that there are no roads to be maintained through the goaf.

The respective advantages of the two kinds of longwall working depend much on circumstances; but the retreating method has the superiority that if in a seam that is liable to spontaneous combustion a gob-fire should break out, the danger is left behind and does not come between the working-face and the shafts.

Gob-fires are brought about through spontaneous combustion of the slack and coaly refuse left behind in the gob or goaf, and are caused probably by the oxidation of the finely-divided carbonaceous matter, assisted maybe by the presence of pyrite. In some cases heat due to pressure and movement has set up spontaneous combustion in the seam itself, as in the case of the Thick Coal at Hamstead in South

Staffordshire, which is being worked at a depth of over 2000 feet.

Comparing bord-and-pillar with longwall work, it may be pointed out that the latter is not applicable to areas underlying water—such as the sea, rivers, lakes, and reservoirs ; nor can the removal of pillars on the bord-and-pillar system be applied to districts occupied at the surface by buildings in respect of which compensation for damage would be demanded. Bord-and-pillar is best applied where the surface must remain supported by the pillars ; longwall yields the largest percentage of large coal, is more easily ventilated, is generally less costly, and is gradually superseding bord-and-pillar.

Square-work.—In working the Thick Coal of South Staffordshire, which varies from 14 to over 30 feet in thickness, and is liable to spontaneous combustion, the method known as Square-work has been evolved. The seam is divided into a number of rectangular compartments, 50 yards or more in the side, called 'sides of work,' separated from each other by ribs of coal (' fire-ribs '), 8 or 10 yards thick (Fig. 14, p. 107). Access to the sides of work is gained by one or more 'bolt-holes,' opened out from the main roads or gate roads cut in the lower layers of the coal. The coal is then removed from within each side of work, leaving a vast gloomy chamber in which six or more pillars remain to support the roof. The pillars are then pared down as far as is safe, and the bolt-holes finally sealed up to prevent spontaneous combustion. In order to get

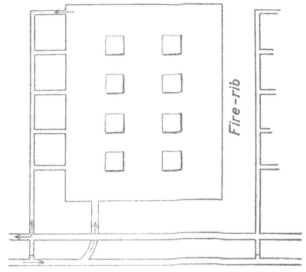

Fig. 14.—Plan showing a ' side of work ' in the Square-work method of getting the Thick Coal of South Staffordshire. The arrows show the direction of the air-current.

at the upper layers of the coal, the hewers stand on the coal and slack already cut, or on light scaffolding. No roof-timbering can be employed, and the work is attended with great danger from falls of roof. Sometimes all the coal is extracted in two lifts or

layers at intervals by a system of longwall work.
But after this first working 38 to 46 per cent. of the
coal is left underground, and a large amount of slack
is produced ; though much of this coal is recovered
when the ribs and pillars are worked by a second
and third working after an interval sufficient to let
the roof settle down.   The removal of the Thick
Coal pillars in South Staffordshire has occasioned
an enormous amount of damage to the surface-
property, often without any compensation being
obtainable by the owners.

Coal-cutting Machines.—The ordinary method of
undercutting the coal with the pick has been described already (p. 98).   The holing in a thin seam
makes a much larger proportion of small coal and
slack than in a thick one.   To remedy this, and to
save time and labour, various types of coal-cutting
machines have been introduced, constructed on the
percussion system, or on the disc, bar, or chain
systems, or on a rotary plan.   In percussion machines,
the cutting-tool is a chisel-ended bar, which is lunged
forward against the coal-face at the rate of 200 or
more blows a minute, rotating at the same time ; and
by swinging the tool slowly along, the hewer cuts a
groove in the coal-face.   The disc-machines, specially
applicable to longwall faces, consist essentially of
a disc or wheel, 5 or 6 feet in diameter, armed with
cutters, somewhat like a circular saw but on a

vertical axle. In the bar-machines, the cutter is
a toothed bar or roller, which is caused to rotate,
at 200 to 500 revolutions per minute, against the
coal-face. The chain-machines differ from the disc-
machines in that the cutters are fixed on an endless
chain, which passes round horizontal pulleys. The
rotary heading-machine, for driving headings in
the coal, consists of a pair of cutting-arms, $4\frac{1}{2}$ feet
long, fixed at right angles to the end of a cross-bar,
the centre of which is attached to a revolving shaft.
As the machine revolves, the two arms scrape out
a circular groove in the coal, forming an internal core
which breaks away in pieces and is removed as fast
as produced. A cylindrical passage 4 to $7\frac{1}{2}$ feet in
diameter is thus cut out. In another similar machine
the cross-bar has no arms, but is fitted with cutters,
which chip out the whole cylinder at once and
produce no large coal.

CHAPTER VIII

VENTILATION, DRAINING AND LIGHTING

Ventilation.—It will readily be gathered that the
atmosphere of a mine speedily becomes vitiated
by the breathing of men and horses, the burning
of lights and explosives, by gases—explosive or

poisonous—given off from the coal or evolved by gob-fires, and by the coal-dust diffused in the air ; and it is essential that this foul air should be continuously and steadily swept out of the mine and replaced by fresh. The fresh air required varies from 100 to 500 cubic feet a minute per person employed in the mine, according to whether the mine is free from gas, or is fiery.

The chief gas evolved from the coal is fire-damp (methane or marsh gas, CH_4), the specific gravity of which, compared with air, is ·559. It was produced probably during the conversion of the original vegetable matter into coal, and remains locked up under pressure within the pores of the coal till liberated in considerable volume when the seam is broken into at the face, whence it may sometimes be heard to issue with a slight hissing sound. Occasionally it jets out from joints and fissures in exceptionally large volumes known as ' blowers ' and ' outbursts.' The gas is colourless, tasteless and odourless, will not support life or combustion, but burns with a blue flame. If present in air to the extent of 9·5 per cent. the mixture will explode violently, producing carbon dioxide (CO_2) and water-vapour, which, with the residual atmospheric nitrogen, forms a mixture incapable of supporting life and constituting the deadly ' after-damp ' of explosions. Being so much lighter than air, fire-damp rises to the higher parts of the workings. It accumulates in the goaves, whence on a fall of the barometer it is liable to issue in dangerous quantities. Its presence in the air of a mine to the extent of only 2 or 3 per cent. causes a blue cap to appear over the flame of the Davy-lamp, the cap increasing in size with the percentage of gas till the lamp is filled, or the mixture explodes.

Carbon dioxide (choke-damp, black damp, stife, CO_2) is produced by the breathing of men and animals and by the complete combustion and slow oxidation of carbon compounds. As we have seen, it is one of the products of the combustion of fire-damp. Being 1·529 times as heavy as air, it accumulates in the lower parts of the mine. It is incombustible, and a non-supporter of life and combustion. Less than 15 per cent. of carbon dioxide in air causes drowsiness when breathed, and in larger quantities the gas is fatal.

Carbon monoxide (sweat-damp, white damp, CO) is produced by the incomplete combustion of carbon compounds (as in gob-fires), by explosives, and by explosions of fire-damp. Its specific gravity is ·967; it is combustible, but will not support combustion. Its worst character is its actively poisonous effect on the blood, so that even 1 per cent. is fatal; and unfortunately this percentage, as it does not affect the combustion of a candle, gives no warning of its presence. Those who have succumbed to it acquire an unnatural ruddiness of the complexion.

Hydrogen sulphide (stink-damp, H_2S) is produced by the decomposition of iron pyrite (FeS_2) in the presence of moisture. Its specific gravity is 1·171; it is combustible, but a non-supporter of combustion or life. Its well-known smell—of rotten eggs—enables it easily to be detected, and even 1 per cent. of it in air is injurious to breathe, though such air will support combustion.

Such then, together with the coal-dust, are the chief impurities that vitiate the air of a coal-mine. The carbon monoxide is the most poisonous; but fire-damp, on account of its abundance and the danger of its exploding, is most to be feared, though as a source of danger coal-dust approaches it very closely. After an explosion, the oxygen essential to respiration has been burnt up more or less completely, to form carbon dioxide, carbon monoxide and steam; so that those who have escaped death by direct shock or burning are cut off by asphyxiation or poisoning. The mechanical effects of an explosion are not only disastrous to the men, but destructive to all impediments in the way of the expanding gases; doors, stops and air-crossings are blown out, tubs are overset, props are thrown down, and the shaft-fittings and winding-gear deranged, with the result that heavy falls of roof are produced and the

ventilating current short-circuited or stopped. Thus
the usual means of clearing the air are rendered
unavailable, and access to the scene of the disaster
may be for a while impossible.

Experiments go to show that air charged with
coal-dust is itself explosive, while as little as 2 per
cent. of fire-damp in air is enough for explosion,
provided that the air is dusty. A dry, dusty and
fiery seam is always dangerous. But by sprinkling
the roads with water, and by using only those explosives which give out as little flame as possible, by
strict attention to the safety-lamps, and by maintaining a generous current of air, these risks may be
greatly reduced. But a sudden blower of gas, which
may for a while overpower the ventilation and render
the atmosphere highly explosive, is always to be
feared, and under the best of circumstances remains,
a grim spectre in the background, ready to leap forth
and deal out death and destruction at the first oppor-
tunity.

Some few mines are still ventilated by the natural
current set up by a difference of density of the air
in the two shafts, usually consequent on their being
of unequal depth, temperature or dryness; but as
such a current, never very powerful, is apt to cease
altogether at changes of the season's temperature,
this natural ventilation is usually replaced by furnaces and exhausting-fans.

Furnaces.—The ventilating-furnace, in which a roaring coal-fire is kept burning, is placed near the bottom of the upcast shaft, so that the latter becomes a huge chimney up which a continuous and powerful draught is maintained. It is insulated from the coal on each side by a brickwork arch and walls. The return air (*i.e.* the foul air returning from the work-ings) passes through, over, and by the side of the fire, and so up the shaft. If however this return air is so charged with fire-damp as to be liable to explosion at the furnace, it is carried up an inclined 'dumb drift' instead of past the fire, and enters the shaft 50 or 60 feet above the bottom; and air sufficient to maintain combustion is led to the furnace direct from the downcast shaft. A furnace should be capable of producing a current of 6000 cubic feet per minute for each foot of breadth of fire-bars.

Fans.—Exhausting-fans are placed on the surface near the top of the upcast shaft, and connected therewith by an air-tight brickwork passage—the shaft-top being closed. They are usually made on the centrifugal principle. As the fan rotates at a high velocity (up to 300 revolutions per minute), the air tends to be thrown out towards the periphery by the vanes, and so produces a low-pressure area round the axis. If therefore this axis of the fan is open to the upcast shaft, and the periphery to the outside air, a continuous current will be set up, and

the air thus sucked up the one shaft will produce a
corresponding influx at the other.   Centrifugal fans,
such as the Guibal, Waddle, Schiele, and Capell fans,
are capable of producing currents of over 200,000
cubic feet a minute.

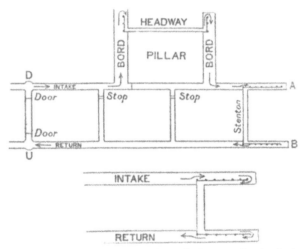

Fig. 15.—Plan showing method of ventilating a pair of winning
    levels and two bords.   Below is an enlarged plan of the ends of
    the levels showing bratticing.   The arrows show the direction
    of the air-current.

Distribution of Air.—The distribution of air in
the mine must now be considered. If left to its
own devices, the air would go direct from the

downcast shaft by the shortest cut to the upcast.   To
prevent this, it is guided by various stoppings and
doors, and not allowed to reach the upcast till it has
gone the whole round of the workings.   Those
passages used for carrying fresh air into the
workings are called the 'intakes,' while those
carrying foul air to the upcast are called the 'returns.'
Fig. 15 (p. 115) will show how the air is compelled
to reach the face in two main roads being driven from
the shafts, and in a pair of bords opening out from
the main roads.   As the cross-heading or stenton
between the two shafts is frequently used as a
travelling road for men and horses, it must be closed
with a pair of tightly-fitting wooden doors, hung so
that they are self-closing, and opening toward the intake.   As the two main roads DA, UB advance, new
stentons are cut at intervals of 30 or 40 yards, and
the old ones, if not required for travelling, are closed
with a permanent stopping of brick or stone.   As
the main roads are advanced beyond a stenton, the
air is conducted up to the face in each case by means
of canvas or wood bratticing (p. 101).   Thus by
driving out the winning headings in pairs or triplets,
a complete air-current can be maintained right up
to the face.   Stone drifts, however, are usually
driven singly, and must be ventilated by a brickwork brattice, or by carrying the air in wooden
pipes (air-boxes) or iron tubes, along which it can

conveniently be forced by small electrically driven fans placed at suitable intervals.

In the various complications of a mine it is frequently necessary to carry one airway across another by what are called air-crossings ; and since the intakes, on account of the freshness of the air, are generally used as travelling ways for the trams, men and horses, it is usual at a crossing to carry the return over the intake by a wooden or brickwork arch.

In early practice it was the custom to carry the whole of the air-current direct to the working-places and back to the upcast ('face-airing'), thus leaving all the old excavations unventilated ; later, it was led in one continuous current through every part of the mine ('coursing the air') ; but this method had the disadvantage that when the last of the working-places was reached the air was already vitiated. The present-day practice is to 'split' the air (see Figs. 12 and 13), at the bottom of the downcast, into several intakes, each of which is taken, direct to its own district or panel, along the main travelling road ; and after each split has done its work it rejoins the others near the bottom of the upcast. To prevent the nearer districts taking more than their share, the current is regulated by sliding doors placed across the returns.

The ventilation of a mine worked by bord-and-pillar is a very complex affair, necessitating a large number of stoppings, doors and brattices (Fig. 12). The air for each district is usually carried up the leading roadway to the most advanced working-place, where it splits and goes right and left; but instead of going back at once by the return it is guided by stoppings and brattices into every working-place and along most of the roadways. The long distance travelled by the air requires a powerful driving-force, and a constant watch against short-circuiting.

In longwall working the ventilation is much simpler and requires fewer stoppings, doors and brattices. The air is carried up the middle gate road to the face, where it splits, travels right and left along the face, and then goes back to the return. The shorter distance travelled by the air needs a less powerful driving-force.

Drainage.—The water almost invariably present in the pervious conglomerates, grits and sandstones of the Coal Measures, down to a depth of 300 or 600 feet, has always constituted a heavy burden on mining enterprise, and in many districts for long periods effectually crushed it. The water so encountered in a pervious bed is of course derived in the first instance from the rain that falls on the outcrop, whence it passes down in the direction of the dip, and is ready to pour into any shaft that reaches that particular stratum. Impervious shales and clays, on the contrary, hold little water, and a coal worked under a cover of such rocks is usually dry. Some of the deepest pits are the dryest, owing to the tubbing-off of the overlying wet strata encountered in sinking (pp. 84–5). Where tubbing has been carried out properly, little water need make its way down the shaft ; but if a water-laden bed is faulted-down in the workings and has to be driven into, it may give rise to a troublesome and even dangerous feeder of water. Similar difficulties may arise through a heavy fall of impervious roof letting the bottom out of some overlying water-logged bed. In working the Shallow Coal at Brereton (South Staffordshire) in 1908 a thin bed of impervious clay, which separated the coal from the water-logged Trias above, suddenly collapsed ; water poured into the mine, drowned several men, and flooded the workings.

A very simple problem of draining is presented by a seam of coal that crops out on relatively high ground, such as the side of a hill (Fig. 1, p. 9), and at the same time dips toward the lower ground. Here the day-level A, by which the coal is worked, will also serve as a drain. But on the western side of the valley in Fig. 1 the workings would soon fill with water. To drain them a drainage-tunnel (adit or sough) is driven in from the bottom of the valley, with the slightest possible upward inclination, till it meets the coal, in which a heading can then be driven level-course till it intersects the slant driven in the coal from the outcrop. The coal can then be worked in such a manner that the workings will drain themselves by the adit. Beyond the adit, however, the limit of free or natural drainage is reached ; and any further workings toward the dip (' dip-workings ') are impossible without resort to artificial means of removing the water.

Only in districts trenched by deep valleys that cut through the coals, such as the Forest of Dene and the Pennant country of South Wales, can such methods be applied ; and in most of such districts the limit of free drainage has long been passed, and the water has all to be raised by shafts.

Water may be raised up the shaft by either winding or pumping. By the first method, a water-tank fitted with an inlet-valve is fixed in or under the cage. The tank is lowered into the water of the sump, where it fills itself by the valve in the bottom. It is then wound to the surface, where a simple self-acting contrivance opens an outlet-valve in the side, upon which the water pours out into a channel ready to receive it. But unless a subsidiary shaft can be turned to account for water-winding, lift-pumps or force-pumps are usually employed.

When dip-workings descend below the level of the shaft-bottom, the water must be conveyed in some

way to the sump before it can be lifted or forced
to the surface by the pumps. A syphon may be
employed where the intervening height over which
the water has to be conveyed does not exceed about
25 feet. Compressed air or steam can be used to
work a pump placed in the lowest workings, or
hydraulic pumps and oil engines may be used ; but
electrically-driven pumps are specially convenient
for this work, and are now usual.

The engine employed to give motion to the pump-
rod of a lift-pump is necessarily placed at the top
of the shaft ; and as has been pointed out (p. 17),
the first steam-engine was constructed for the express
purpose of pumping water from mines. The old
Cornish pumping-engine as developed by Watt, with
its complicated mechanism and huge beam, from
the end of which depended the pump-rod, is now
seldom seen ; and the water in a modern colliery is
usually raised by force-pumps placed in the workings
and supplied with steam generated at the surface,
though electric centrifugal pumps are now installed
at many mines. In parts of the South Staffordshire
Coalfield a system of general drainage has been
established to unwater the mines. Large pumping-
stations have been set up, and are maintained by a
charge of a few pence per ton levied on all the coal
raised at all the collieries in the district.

Lighting.—The methods to be adopted for lighting

the mine depend on the presence or absence of fire-damp. In its absence, naked lights can be employed. At the shaft-bottom or sidings, where there is much traffic and a good light is necessary, gas-jets or large oil-lamps are convenient ; for travelling along the main roads portable oil-lanterns can be used; while at the working-face tallow candles are handy, as they can be set upright in a lump of clay placed in any position required. In mines where fire-damp is encountered, naked lights may be admissible at the downcast shaft-bottom, and for some distance along the intake airways, but at a certain point must be exchanged for safety-lamps.

The principle of the safety-lamp (p. 31) depends on the well-known fact that a fine wire gauze if kept cool will reduce the temperature of a flame to a point below that necessary for combustion. Davy applied this principle in his safety-lamp, which consisted of an oil lamp closely surrounded by a gauze cylinder 6 inches high and $1\frac{1}{2}$ or 2 inches in diameter, and closed at the top with a gauze lid. If such a lamp be placed in a mixture of fire-damp and air, a blue cap appears over the flame and increases in size with the percentage of fire-damp present ; but as long as the flame does not make the gauze red-hot, and the lamp is not exposed to a current of such air having a velocity greater than 6 feet per second, the flame burns safely within the lamp and does not ignite the explosive atmosphere without. With an excessive amount of fire-damp present, however, the whole lamp is filled with burning gas; the gauze becomes red-hot, and the flame will pass through and ignite the mixture outside.

Since the date of Davy's invention in 1815, numerous lamps based on the same principle have been introduced, and designed to give a better light and to be safe in the rapid air-currents maintained in modern ventilation. Many of these, such as the Bonneted Mueseler, the Marsaut, and the Tin Can Davy, are safe in a current of 40 feet per second. In illuminating-power, however, even the best do not attain to half a candle-power.

The safety-lamp supplies in itself a means of detecting the presence of fire-damp in the air of the mine (p. 122); but special fire-damp detectors of great delicacy are now used, the most satisfactory of which are based on the same principle.

Electric light is now largely applied in the surface-buildings and at the shaft-bottom; but in the underground roadways and at the working-face, where falls of roof are liable to damage the insulation, it is risky, and portable incandescent lamps furnished with a small storage battery have been introduced to meet the case. The proper cleaning, trimming, examination and testing of the safety-lamps form a very important part of the colliery management's duty.

CHAPTER IX

UNDERGROUND HAULAGE, WINDING, AND SURFACE-ARRANGEMENTS

Underground Haulage.—At the working-face the coal is loaded into small wooden or sheet-iron trucks running on four wheels and known as tubs or trams, their capacity varying from 5 to 20 cwt. Each tub is furnished with a few links of chain at one end and a hook at the other. The wheels are flanged on their inner edge as in an ordinary railway truck, and run on iron rails of small gauge. The underground railways are of various degrees of permanency and solidity. At the face they are light and temporary, so as to admit of being removed and laid down again with ease as the face advances. Along these temporary railways the tubs are pushed by 'putters' or ' trammers ' to a convenient point known as a siding, flat or station, where the railway from the face joins a more important road called the horse-road or rolley-way. Here the tubs are made up into a train, which, if the distance is short and the road fairly level, is then drawn to the shaft-bottom by horses. But if the distance is great or the gradient heavy—either down- or up-hill—mechanical haulage is employed. If the gradient outbye has a moderate dip, say an inch in four yards, horses can

pull the loaded tubs down to the shaft; if much
steeper, the tubs would over-run the horses, and the
latter can be dispensed with. In such a case the
loaded tubs can pull up the empty ones by a self-
acting incline. At the top of a straight road leading to the shaft is fixed a drum or pulley furnished
with a brake; round this is passed a rope, to one
end of which is attached a loaded train and to the
other an empty train. The loaded train being
started at the top of the incline descends and pulls
up the empties—the pace being regulated by the
brake. Two sets of rails can be employed where
the road is wide, but otherwise a single line, with a
length of double rail at the meeting-place, where the
trains pass each other, will suffice. By means of a
small easily-moved pulley and brake the same system is employed, at a working-face advancing to the
rise, to run several loaded tubs down an incline called
a gig-brow to a siding or main road.

In many collieries, however, coal has to be raised
from dip-workings that lie far below the level of the
shaft-bottom, where gravity can be employed only
partially; or the working-places, although on much
the same level as the shaft-bottom, are reached by
roads that undulate up and down hill; or the workings may be situated long distances from the shaft.
In such cases horses are inadequate, and a system
of mechanical haulage must be installed. For this

purpose stationary engines are employed, either on
the surface or underground. Where on the surface,
the ropes are carried down the shaft in a wooden
casing ; where underground, the engines are supplied with steam generated at the surface and carried
down the shaft in pipes. Several different haulage
systems have been developed.

The ' main-rope ' or ' direct haulage ' system, for
raising coal from dip-workings, can be employed
where the gradient of the engine-plane (*i.e.* the inclined plane up which the tubs are hauled from the
workings to the shaft-bottom) has a regular dip of
not less than $1\frac{1}{2}$ inches in a yard. At the station
a train of loaded tubs is hitched on to the end
of the rope ; a signal is given to the engine-room
on the surface, the rope is wound in, and the train
pulled up to the shaft-bottom. There the rope
is unhitched, attached to an empty train, which is
then started down the incline. The gradient being
sufficiently steep, the train pulls in the rope with it,
the drum at the engine meanwhile being thrown out
of gear so as to run free on its shaft. One drum,
one rope, and one set of rails suffice for this system.

The ' main-and-tail-rope ' system (*A*, Fig. 16,
p. 127) is applicable to almost any condition of road.
Only one set of rails is required, but two parallel
ropes are needed. One, the main rope, hauls out
the full tubs ; the other, of lighter construction,

hauls in the empties. At the engine-house each rope is wound on a separate drum, which can be thrown out of gear as required. At the inbye station the end of the tail rope is passed round a return pulley. In Fig. 16 the full train is being hauled outbye by the main rope, whose drum is in gear ; at

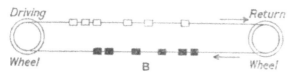

Fig. 16.—Plan showing Main-and-Tail-Rope system of haulage (*A*) and Endless-Rope system (*B*).

the same time the tail rope is being drawn off its own drum, which is running free. At the shaft-bottom, the full train is replaced by an empty one, which is hauled inbye by the tail rope, while the main rope is being drawn off its own drum, now out of gear. The system is capable of extension to branch-roads in several different ways.

The ' endless-rope ' system (*B*, Fig. 16) requires

two sets of rails, one for empties travelling inbye, the other for full tubs going outbye. At the shaft-bottom the single rope is carried round a driving-pulley actuated by the engine, while at the inbye station it passes round a return-pulley. The tubs are attached to the rope singly or in short trains of two or three, by means of clips, and the rope travels continuously, either above or beneath the tubs. In order to insure that the rope is kept taut, it is passed round a pulley fixed on a movable tram to which a hanging weight is attached. The system can be extended to branch-roads by causing the main end-less rope to give motion by suitable pulleys to a separate rope for the branch.

Winding.—The coal having arrived at the bottom of the winding-shaft, the next procedure is to raise it to the surface. For shallow shafts not exceeding 30 yards or thereabouts, or for small trial-shafts, windlasses of various powers may be used, with one or two buckets, and a hempen rope. For greater depths and heavier loads a horse-gin is sometimes still employed, as at small collieries in South Staffordshire. Usually, however, the coal-tubs are placed in a cage running between guides, and raised to the surface by a steel rope wound by powerful steam-engines. The winding-shaft generally accom-modates two cages, one ascending with full tubs while the other descends with empties.

The cage is the receptacle that carries the tubs, men and materials up and down the shaft. It is constructed of iron or steel, and has one to four decks, and takes one, two or three tubs on each deck. Each deck is furnished with rails on which the tubs stand, and each tub is kept in position by a catch made to grip its edge or its axle. The cage is supplied with forked or tubular slides, which engage with the guides carried up the shaft-sides, so that while the cage is travelling up and down—even at such speeds as 80 feet a second—all swinging, spinning, and bumping, against either the sides or the other cage, are eliminated. The cage is suspended to the rope by wrought-iron chains.

The guides are strong wooden, steel or iron rails, or steel ropes, placed vertically in the shaft for the smooth guidance of the cage. Their number and position will be determined by the form and size of the cage, and the kind of guide used. Wooden guides are made of rails of pine-wood, 18 to 20 feet long, and 4 by 3 inches in section, placed end to end and bolted to horizontal cross-pieces (buntons) fixed to the shaft-sides. Iron or steel rails make a more substantial and durable guide, and wire ropes also are largely employed, fixed at the top to the pit-head frame, passed at the bottom through balks, and kept taut by heavy weights hung to their lower ends.

The ropes used for winding are usually round in section, and made of steel wire. The mouth of the shaft is fenced with iron or wooden gates, which are automatically lifted up by the cage when it reaches the top landing (and so takes their place), and fall into position again as the cage descends. The cage is kept in position, flush with the landing-place, by 'keeps,' which by an arrangement of levers are shot forward under the cage when it reaches the top.

Signals are communicated between the engine-house and the top and bottom landing-places, and also underground, by a wire, actuated by a lever, causing a rapper to strike a sonorous iron plate; but electric bells and telephones are now becoming general.

At the top of the winding-shaft the winding-ropes are carried over pulleys hung in the pulley-frame or pit-head frame, and thence enter the engine-house. Pulley-frames are built of wood or iron—the latter is more durable and obviates risk of fire. Two or four uprights (pulley-legs) 30 to 80 feet in height are erected vertically over the shaft, and two back-stays are placed at a slope from the top of the pulley-legs so as to resist the pull toward the engine-house. Where, as is usual, two cages are employed in the same shaft, the pulleys are placed side by side in the pulley-frame. The pulleys are grooved to suit the diameter of the rope, and are 10 to 20 feet in diameter, dependent on the thickness of the ropes. Small pulleys involve a sharper bend in the rope, and so increase its liability to fracture.

It occasionally happens that the rope breaks, or the cage, instead of being stopped at the landing-place at the shaft-top, is ' over-wound,' *i.e.* pulled up to the pulleys, when the rope breaks and the cage —with its mineral or human freight—falls down the shaft. Various devices have been introduced to prevent these calamities. Safety-cages are provided with an arrangement that strongly grips the guides as soon as the rope breaks, and so keeps the cage suspended. In the event of overwinding, the rope is liberated from the cage, and the latter is prevented from falling, by the insertion of a detaching-hook between the end of the rope and the cage-chains. Just below the pulley the rope passes through a bell-shaped socket let into a wooden cross-beam or catch-plate. When the cage is inadvertently pulled up to the catch-plate, the jaws of the detaching-hook are automatically opened, the rope goes free, and at the same instant two catches spring out above the catch-plate and keep the cage suspended.

Another method of preventing overwinding is to apply a powerful steam brake to the winding-drum through the action of levers placed in the path of the cage at a point above which it ought not to pass when under control.

The most approved form of winding-engine is the horizontal direct-action coupled engine, which has two horizontal cylinders, with connecting-rods attached to cranks on the axle of the winding-drum. The cranks are set at right angles to each other, so that there is no dead-centre; the drum is provided with band-brakes, actuated by the foot of the engine-driver, or by steam. The drum is cylindrical, with high flanges to prevent the rope slipping off; and the rope is so wound upon it that when one cage is descending, the other is ascending. In order that the engineman may bring the cages to a standstill at the landing-places, it is essential that he should see at a glance the position of the cages in the shaft. This he can do by the winding-indicator, which consists of either a pointer travelling in a vertical frame and actuated from the drum-shaft, or a dial on which a pointer indicates the cage's position. Usually a bell is caused to ring in the engine-house as the cage approaches the top landing, and it is important that the engineman while at his post should have a clear and uninterrupted view of the shaft-mouth.

With a deep shaft involving a great weight of rope it is obvious that there will be a much heavier strain on the engine at the commencement than at the end of the lift. In order to assist the engine at the beginning of its wind, various methods of counterbalancing have been devised.

Surface-Arrangements.—The top landing-stage, where the tubs are withdrawn from the cage and replaced by empty ones, is not laid out on the ground-surface, but is raised 20 or 30 feet above it on an earthen bank, or, still better, on iron pillars. The height thus gained allows the tubs to be emptied over screens into the railway-trucks brought up alongside or underneath to receive the coal. The landing-stage is generally paved with iron plates, on which the tubs can be turned and moved about with much facility, gravity being utilized wherever possible to help in running the full tubs toward the screens.

The full tubs, after being weighed, are emptied on to the screens, not by laboriously shovelling out the coal, but by running the tub into a tippler, which consists of an iron framework rotating on a horizontal axle, and placed over the screen. By a suitable movement the tippler is rotated till the tub is turned over sufficiently for the coal to shoot out. The screens or riddles are required for the purpose of sorting-out the coal into various-sized pieces. Coal in large pieces—called ' round ' coal, for no obvious reason—commands a high price, while the fine dust or ' duff ' can scarcely be got rid of at any price except for the manufacture of patent fuels. Intermediate sizes are known by such names as cobbles, nuts, beans, peas, etc. The ordinary screen

is a rectangular iron spout, in the flat bottom of
which are fitted gratings made of iron bars placed
lengthwise. The screen is set with an inclination
of 20° or so, and the bars are fixed at a suitable
distance apart to allow all but large coal to fall
through into a hopper as the contents of the tub
slide down the screen. The large coal that does
not so pass through lands on a table, where any
lumps of bad coal, bat, stone or pyrite are picked out
by hand and thrown on one side. The coal is then
shovelled into the truck standing before the table.
During all these operations the coal is bound to suffer
further breakage, and a soft coal suffers seriously.

It is necessary to subject some varieties of coal
to a much more elaborate screening and cleaning,
in which case shaking-screens and travelling belts
are employed. Shaking-screens are suspended at a
small angle and caused to jerk backward and forward about 70 times a minute by a special engine.
The coal travels slowly down-hill toward the truck,
while lads stand alongside and pick out the rubbish.
At the bottom of the screen the coal is delivered
on to a horizontal travelling belt, 20 to 60 feet long
and 3 or 4 feet wide, made of iron plates or
cloth, and caused to travel slowly forward. The
coal, evenly and thinly scattered over the belt, then
passes under the scrutiny of the hand-pickers, who
throw out any refuse still undetected.

Small coal destined for the coke-manufacturer has generally to be washed to free it from impurities. The principle on which coal-washing machines are constructed is that the specific gravity of coal is smaller than that of its stony impurities. When impure coal is agitated in a stream of water, the heavy impurities move toward one part of the machine, while the coal, being lighter, moves toward another. Some of the larger washers will deal with hundreds of tons a day.

In working the coal, much rubbish (bad coal, pyrite, shale, and stone) is produced, and must be disposed of. Some is stowed underground in the goaf; but much of it is brought to the surface and piled up in huge unsightly mounds, where it frequently takes fire spontaneously, filling the air with noisome vapours. When burnt out, however, some of these rubbish-tips are turned to account for ballast, the burnt shale, which often assumes a brick-red colour, being specially suitable for garden footpaths. In South Staffordshire many old tips have been successfully planted with timber, and converted into picturesque and attractive recreation-grounds, as at Wednesbury and Walsall.

CHAPTER X

LEASES AND ROYALTIES, ADMINISTRATION, AND STATE REGULATIONS

Leases.—Though in early times all minerals in the British Isles appear to have been claimed by the Crown, this claim has unfortunately long been relinquished except in the case of gold and silver ; and in the absence of other provisions the minerals belong now to the owner of the surface. A property may, however, be sold subject to the minerals being reserved, or subject to the right of the vendor to work the minerals on paying compensation for surface-damage. A mining-company may purchase the minerals from the owner, who will then retain possession of the surface. The owner of a property may work the minerals himself, but usually he retains his surface-rights and grants a lease of the minerals to a mining-company. Under ordinary custom of mining, the lessees are liable for all surface-damage. The terms on which the mineral-owner will grant a lease vary greatly according to local custom. Leases are granted generally for 21, 42, or 63 years, with power of surrender.

As a general rule, colliery-proprietors show great reluctance to expend capital in purchasing mineral properties out and out, even when they can do so upon highly-favourable terms; and the vast majority of collieries are leased by the colliery-owner from the owner of the soil. The terms of these leases vary in different districts; but the general features have become standardized as the result of generations of negotiation and experience.

For the surface-land necessary for the erection of the pit-head machinery, railway-sidings, and for tipping-ground for rubbish, a fixed rent per acre is charged—in South Wales the usual rate is £2 per acre.

Dead-rent and Royalties.—Then there is a definite annual sum payable for the whole taking, known as the 'fixed,' 'sleeping,' 'minimum' or 'dead' rent. This is roughly calculated upon the total area of the mineral taking, and its primary object is to insure that the lessee will actually work the minerals and not merely hold them up.

In the case of a colliery that is working fully, the amount of the dead-rent is not very material, as it merges into the 'galeage' or 'royalty,' which is the essential feature of a mining-lease of coal. This royalty is a payment to the landlord proportionate to the amount of coal worked from the mine during a certain period (usually a year), and the method of calculating it varies considerably in different districts. One method that is very common in the case of metalliferous mines, and is sometimes adopted for coal also, is to give the landlord a proportion (say $\frac{1}{14}$) of the actual selling-price of the coal at the pit-head. Another method (the Yorkshire custom) is called the 'acreage royalty.' In this system the actual area of the coal worked is calculated at the end of the selected period, and payment is made at the rate of so much per acre upon the coal worked. The rate is either a simple one per acre, or it may vary, according to a sliding-scale, with the actual selling-price of the coal. A modification of this method (the Nottingham custom) is the 'footage royalty.' In this case, instead of calculating the area simply, without regard to the thickness of the seam, the unit taken is an acre of coal one foot thick (the 'foot-acre'), so that if an acre has been worked of a seam two feet thick, two footage royalties would be payable; but if the seam were ten feet thick, ten such royalties would be payable, and so in proportion. Otherwise it is calculated in the same way as the acreage royalty, either simply, or upon a sliding-scale.

The third method, customary in South Wales and Durham, is the 'tonnage royalty.' In this case the royalty is paid at so much per ton (say $3d.$ to $10d.$) upon the actual quantity of coal worked—the rate varying according to the quality and thickness of the seam, and sometimes also varying in proportion to the selling-price of the coal.

Whatever method may be adopted, however, the royalty always merges in the dead-rent; so that the royalty is only payable if the amount of it exceeds the amount of the dead-rent. Further than this : almost all mining leases contain some form of average-clause. Under this, if in one year the mine-owner has failed to work such a quantity of coal as would make up the dead-rent in royalties, but in the succeeding years over an agreed period the royalties exceed the dead-rent, the excess is set off against the deficiency.

Suppose, for example, that the dead-rent under a particular lease is £600 a year, and the average period three years; also that in the first year the royalties amount to £300, in the second year £400, and in the third year £500 ; the total royalties are £1200, and the average yearly royalties £400. Therefore as the average royalties do not exceed the dead-rent, £600 a year is payable to the owner, who has thus made at the end of the three years £600 more than the earnings of the colliery have yielded !

But suppose the dead-rent is £600, the average period three years, and the royalties are respectively £500, £700, and £900. The total royalties are £2100, and the average £700. Therefore as the average royalties exceed the dead-rent, the royalties are payable, *i.e.* £2100. If there had been no average clause, the dead-rent of £600 would have been payable in the first year, and the royalties in the next two years, or £2200 in all as against £2100.

Mining leases usually contain also a provision enabling the lessee (but not the landlord) to terminate the lease if the coal has all been worked out or has become unworkable at a profit ; and there are very many other provisions that we have no space to mention. It is of interest to know that a Royal Commission reported in 1893 that any reduction in royalties would benefit nobody but the consumer !

Way-leave.—In the lease provision may need to be made for surface way-leave, underground way-leave, and other privileges. Way-leave is the privilege of a lessee to carry his minerals over or through the property of another owner, and is usually paid for by a tonnage on the minerals so carried. Surface way-leave may be a source of considerable profit to the owner of the property crossed, as it sometimes takes the form of a heavy annual charge, fixed in some cases as high as £250 per mile, plus a rent estimated at double the agricultural value of the land required ! Mining in a district much cut-up among small owners, as in parts of South Staffordshire, is subject to endless burdens and inconveniences from the way-leaves, which in the past have entailed a maximum of expenditure and waste in the sinking of superfluous shafts and the leaving of barriers of valuable coal between the several properties.

Administration.—The running of the colliery is placed in the hands of the colliery manager, who nowadays has to be a man of wide and varied knowledge, experience, and organizing power. To get the largest proportion of available coal in the best selling condition, at the lowest cost, and with the greatest amount of safety and comfort to the employed, he must be a well-trained mining-engineer, a good man of business, and a tactful leader of men. For all that goes on in the mine he is responsible, not only to the lessees of the minerals, but also to the State. The most important officials working under him include the under-manager, the overman, deputy-overmen, master-shifter, master-wasteman, engineer, heap-keeper, hewers, fillers, putters, stonemen, shifters, wastemen, banksman, onsetters, rolleywaymen, horse-keepers, furnacemen, trappers, etc. Associated with them is the lessor's agent, who acts for the lessor, looks after his interests, and sees that the conditions of the lease are fulfilled ; also the check-weigher, who is appointed by the men paid according to the weight of mineral obtained, to see that they receive justice.

The under-manager is responsible for the proper running of the mine during the temporary absence of the manager. The overman has responsible charge underground, and makes out the wage-bills ; each district in the mine is in charge of a deputy-overman. The master-shifter is in charge during his shift in the absence of the overman; the master-wasteman looks after the ventilation, safety-lamps, etc.; the engineer is responsible for the engines, boilers and machinery. The heap-keeper superintends the pit-bank, screens, loading of rail-way-trucks, etc.; the banksman has control of the shaft-top, and the onsetters of the shaft-bottom. Hewers are the men who do the actual getting of the coal at the face; fillers place the coal into the tubs; putters convey the loaded tubs from the face to the putters' flat or siding. Stonemen do the 'dead' work of driving or enlarging the stone-headings, and perform excavation-work other than in the coal. Shifters do the necessary clearing away of falls of roof, and set timber in the road-ways; wastemen do the same work in the return air-ways; rolleyway-men see that the underground railways are in order. The trappers are boys who look after the trap-doors in the air-ways and see that they are kept closed after use, as any neglect might disarrange the ventilation and cause an explosion.

Wages and Output.—These various officials are paid in different ways, some by the day (datallers), others, such as the hewers, putters and stonemen, are paid by piece-work. Hewers generally are paid by the ton of coal sent up—deductions being made for a tub not properly filled, or which contains too much rubbish. The hewing-price per ton varies very greatly, and is modified by all sorts of special circumstances. Putters are paid according to the number of full tubs they convey from the face to the flat, and the distance involved. Stonemen are paid by the cubic contents and character of the material excavated, or else by contract.

The men work in shifts, two or three in the 24 hours, of 8 hours each, bank to bank ; and wages are paid once a fortnight.

Taking into consideration eleven different representative collieries in various parts of England and Wales during the year 1904, Messrs Bulman and Redmayne find that the cost of labour averaged 3s. 11½d. a ton for underground men, and 7½d. for the surface-men, or altogether 4s. 7d. per ton, which is only 50 or 60 per cent. of the total cost of getting, the other percentage going to materials, royalty, rates and taxes, and management. The average production per hewer per shift was about three tons ; per underground hand 1·39 tons, and including surface hands, 1·19 tons. The hewers received an average wage of 6s. to 7s. a shift, the other workers earning lower wages, except stonemen, etc., who are paid by the piece at higher wages.

The following recent figures for a Yorkshire mine will give some idea of the scale on which a modern colliery is conducted :

Output per fortnight of 10 days, 18,966 tons, from 6 seams (house-coal and coking-coal) worked by longwall.

Stock of tubs, 3500.   Men employed at surface, 500; underground, 1394.   Shift, 8 hours.

Details of underground labour:

	Number	Percentage	Net earnings per shift s. d.
Hewers	605	43	6 11½
Stonemen and shifters ..	103	7	8 1
Putters	316	23	5 9¼
Drivers and boys	243	18	3 3½
Deputies	18	1	7 3
Various	109	8	5 10¼
	1394	100	

In 1800 the estimated coal-output of the United Kingdom was 10 million tons ; in 1903 it was 230 million tons ; in 1912, 260 millions ; and the findings of the Royal Commission of 1901 show that British coal is not likely to last more than another 250 years, and will become exhausted between A.D. 2130 and 2200.

Accidents.—Although the number of fatal accidents in our coal-mines is still sadly large, it is steadily becoming smaller in proportion to the number of men employed. Occasionally some catastrophe of unusual magnitude, generally an explosion—such as that in 1866 at the Oaks Colliery in Yorkshire, where 334 men were killed outright ; or that at Senghenydd in Glamorgan in 1913, which killed 439; or an irruption of the sea or a river into the workings, such as took place at Landshipping in Pembrokeshire in 1844 and killed 40—stirs the public interest, and calls attention to the perils that must ever attend this dangerous calling. Yet as a fact the death-rate among colliers is kept up by less sensational accidents, such as falls of roofs and sides, or those connected with underground haulage, or in shafts, and from other minor causes, all of which are in operation daily, and claim their victims singly or in small numbers.

In 1912, falls of rock from roof and sides accounted for 44·5 per cent. of the deaths, miscellaneous accidents for 26·5, accidents on the surface for 13·8, in shafts 5·6, and explosions of fire-damp and coal-dust 9·6. In the same year there were employed at the 3093 mines 1,072,393 persons (of whom 6457 were women and girls) above and below ground, and the number of fatalities amounted to 1258.

Explosions of fire-damp, or of coal-dust, or of a mixture of the two, may take place, in spite of a good ventilating-current, through a fall in the atmospheric pressure leading to a rise in the per-centage of fire-damp (p. 111); through a sudden blower or outburst of gas locally rendering the air explosive, when the defective condition of a safety-lamp, or some damage to it, or some careless or reckless conduct on the part of a workman, or the flame of the explosive used to bring down the coal, may lead to a disaster.

Falls of roof and sides may be guarded against by abundant and properly-set timbering. Haulage-accidents arise from tubs breaking loose and running amok down an incline, or through men being knocked down by moving tubs; while in shafts the victim may fall from a side-opening or from the surface, or may be killed through overwinding, or by the breaking of a rope. Blasting-operations too claim their toll of lives, while on the surface accidents happen on the railways and tramlines. The introduction of electricity brings with it new risks of shock to the men and of ignition of fire-damp and combustible materials.

State Regulations.—The first Act of Parliament dealing specially with mine-regulation was passed in 1842, to exclude women and girls from underground employment, to improve the status of the boys, and to provide for the appointment of inspectors. In 1860 an act introduced general rules for working, prescribed certain mechanical appliances, prohibited some dangerous practices, and authorized special rules defining the duties of the officials and workmen—all with a view to the safety of the staff. In 1865 two shafts or outlets to the mine were made obligatory. In 1887 the earlier acts were repealed, and their provisions re-defined and amplified in the principal act dealing with coal-mines, which applies to all mines in Great Britain and Ireland working coal, stratified ironstone, shale and fireclay. Several later acts relate to check-weighers, explosives, employment of boys below-ground, qualifications of managers and under-managers, timbering, and electric installations. The most important of the later acts is that of 1908, known as the Eight Hours Act, while the various Truck Acts, and the Workmen's Compensation Act of 1906, apply to the management of collieries.

Plans and Records.—A most important provision of the Act of 1887 is that at every colliery-office a plan must be kept, showing the extent of the workings up to date, with the direction and amount of the dip, together with a record of the shaft-section whenever obtainable, or at least a statement of the depth, with a section of the seam worked. By the Act of 1896, on the abandoning of a mine a copy of the plan must be deposited at the Home Office, where after a lapse of ten years it becomes available for public inspection. Although of course plans—and sometimes excellent ones—were kept long before 1887, it was not made compulsory till that year; the result being that there are many disused collieries of which no plans are known to be extant. Thus not only has much valuable scientific information been lost to posterity, but many serious practical

10—2

difficulties have been put in the way of later projectors.   In the absence of accurate plans of abandoned collieries, there is a constant danger of the
present-day workings suddenly tapping water or gas
from old works.   Old plans, even where accessible,
are often inaccurate, and in many particulars are
unintelligible ; and in some cases the pits to which
they relate cannot be located for want of the necessary topographical particulars.   Faults are marked
' riser ' or ' dipper ' (upthrow or downthrow) without
any indication of the direction and amount of the
downthrow.   The magnetic north point (a variable
quantity) is given, without the date necessary to
fix the true north point.   On a plan showing several
different shafts, a shaft-section may be given, but
no means of ascertaining to which of the shafts it
refers.   From these and such-like defects, many of
the older plans yield disappointingly meagre information.

Another source of valuable geological information,
viz. the results of the numerous boreholes put down
year after year all over the country, is being still
pitiably neglected by the State.   A boring in search
of coal may be carried out at great cost ; if successful, and a shaft is sunk and coal worked, the case
would come under the provision of the Act of 1887,
and the section to some extent would be recorded
on the plan.   But if no shaft is sunk, or no coal

worked, the information obtained from the boring may never be put on permanent record, unless it should happen to attract the attention of some geological enthusiast, who may—or may not—publish the results in the ' proceedings ' of some scientific society. There seems no valid reason why the State provision as to shaft-sections should not be extended to the case of boreholes, and a record of the strata deposited at the Home Office.

BIBLIOGRAPHY

The foregoing pages provide merely an introduction to the subject of Coal Mining; for further information the reader should consult the following works:

General Geology:
> Any of the modern text-books.

Natural History of Coal:
> E. A. NEWELL ARBER, 'The Natural History of Coal,' 1911 (Cambridge Manuals of Science and Literature).

Fossil Plants of the Coal Measures:
> E. A. NEWELL ARBER, ' Fossil Plants,' with 60 photographs, 1909 (Gowans & Gray).
>
> D. H. SCOTT, 'The Evolution of Plants' (Williams & Norgate, 'The Home University Library').
>
> A. C. SEWARD, 'Fossil Plants,' Vol. I, 1898, Vol. II, 1910 (Cambridge Univ. Press).

Geology of the Coal Measures:
> WALCOT GIBSON, 'The Geology of Coal and Coal-Mining,' 1908 (Arnold).

History of Coal Mining:
> R. L. GALLOWAY, 'Annals of Coal Mining and the Coal Trade,' series 1, 1898; series 2, 1904 (Colliery Guardian Co.).
>
> R. L. GALLOWAY, 'A History of Coal Mining in Great Britain,' 1882 (Macmillan).

Coal Mining :
> W. S. BOULTON (Editor), ' Practical Coal-Mining,' Six Vols,
> 1907–1909 (Gresham Publishing Co.).
> H. F. BULMAN and R. A. S. REDMAYNE, ' Colliery Working
> and Management,' 1906 (Crosby Lockwood & Son).
> C. PAMELY, ' The Colliery Manager's Handbook,' ed. 5,
> 1904 (*Ibid.*).
> R. PEEL, ' An Elementary Text-book of Coal Mining,' ed. 16,
> 1911 (Blackie & Son).

Statistics :
> ' Mines and Quarries: General Report, with Statistics, for
> 1912 ' (Home Office).

Many of the early forms of atmospheric and steam-engines
(both stationary and locomotive), and other mechanical appliances closely connected with the subject of Coal Mining, are
exhibited in the Victoria and Albert Museum, South Kensington,
and are described in the official guides.

INDEX

Milton Keynes UK
Ingram Content Group UK Ltd.
UKHW032321161024
449665UK00001B/13

9 781107 605817